Nasa History

The History and Legacy of Nasa Missions

(The History of the Nasa Programs That Led to the Successful Apollo Missions)

Jeremy Knapp

Published By **Daniel Kaplan**

Jeremy Knapp

All Rights Reserved

Nasa History: The History and Legacy of Nasa Missions (The History of the Nasa Programs That Led to the Successful Apollo Missions)

ISBN 978-1-9994523-8-4

No part of this guidebook shall be reproduced in any form without permission in writing from the publisher except in the case of brief quotations embodied in critical articles or reviews.

Legal & Disclaimer

The information contained in this book is not designed to replace or take the place of any form of medicine or professional medical advice. The information in this book has been provided for educational & entertainment purposes only.

The information contained in this book has been compiled from sources deemed reliable, and it is accurate to the best of the Author's knowledge; however, the Author cannot guarantee its accuracy and validity and cannot be held liable for any errors or omissions. Changes are periodically made to this book. You must consult your doctor or get professional medical advice before using any of the suggested remedies, techniques, or information in this book.

Upon using the information contained in this book, you agree to hold harmless the Author from and against any damages, costs, and expenses, including any legal fees potentially resulting from the application of any of the information provided by this guide. This disclaimer applies to any damages or injury caused by the use and application, whether directly or indirectly, of any advice or information presented, whether for breach of contract, tort, negligence, personal injury, criminal intent, or under any other cause of action.

You agree to accept all risks of using the information presented inside this book. You need to consult a professional medical practitioner in order to ensure you are both able and healthy enough to participate in this program.

Table Of Contents

Chapter 1: The First Pioneers 1

Chapter 2: Rocket From Thor-Able 20

Chapter 3: The Artistry Of Pioneer 45

Chapter 4: The Infrared Radiometer 63

Chapter 5: Venus Projects 100

Chapter 6: The Lure Of Mars 110

Chapter 7: The History Of Nasa 129

Chapter 8: A Few Facts On Artemis Fun Facts 140

Chapter 9: Nasa Project Artemis 152

Chapter 10: Nasa's Ten Very Best Achievements 158

Chapter 1: The First Pioneers

From the very beginning the first people looked towards the night sky and thought about the bright points of illumination. Prior to the advent of urbanization, the night sky was always dim and filled with stars in such a quantity that they were impossible to count. They all were in the Milky Way galaxy, a perpetual cloud that contained million of stars.

William Gilbert (1544-1603), who realized the fact that Earth has two poles in magnetic field, conceived that distant worlds orbiting stars could have their own magnetic fields, and maybe even existence,. In 1610 the famous Italian researcher Galileo Galilei pointed his telescope towards Jupiter and observed there were four moons which revolved around the planet in the same way like the planets around the Sun. Based on distance from the planet they're Io, Europa, Ganymede and Callisto.

It wasn't until nearly 45 years further on when Dutch Astronomer Christiaan Huygens discovered an enormous moon that was orbiting the ringed planet Saturn. On the 25th of March 1655 Huygens was able to discover Titan, one of the moons in the Solar System with a thick atmosphere. Actually, the atmosphere is able to sustain a pressure 1.45 times greater than the pressure at sea levels on Earth as well as Titan is a dimension of 5,150 kilometers (1.48 times larger than the Moon) that makes Titan the second-largest natural satellite of the Solar System after Jupiter's Ganymede. Both Titan as well as Ganymede are greater than Mercury.

Huygens in trust of his own memory, made holes of various dimensions drilled into a screen aimed at to the sun. He found one closest to the brightness apparent of Sirius the Dog Star which is which is the most bright star of the stars in nighttime. Then, he calculated the distance of this star. If

Sirius and the Sun and Sirius were intrinsically similar in brightness in the exact same direction the scientist could have been very close to his calculation. But, Sirius is an A1V bright blue-white dwarf that is much brighter than Sun that is the G2V, the yellow white dwarf star. Sirius according to the current accuracy of measurement is 22.3 times brighter than the Sun and, if they were placed next to each other in the normal 10-parsec distance (32.6 lighting year), Sirius would be 3.37 magnitudes brighter.

On the 13th of March 1781, the German-born English Astronomer William Herschel discovered the planet Uranus. At first, he believed Uranus was a comet nevertheless, astronomers from other fields calculated the orbit of Uranus and discovered that it was nearly circular like other planets.

In the year 1801, Italian priest Giuseppe Piazzi discovered a planet that was orbiting within Mars as well as Jupiter. It was tiny, with a radius of 946 km (27 percent of the moon's size) in addition, Piazzi recommended the name "Cerere Ferdinandea," but the world-wide community of astronomers decided on Ceres. Presently, Ceres is referred to as"a "dwarf Planet."

In the following the year German Astronomer Heinrich Wilhelm Olbers discovered a second minor planet located between Mars and Jupiter and is today known as Pallas. The average diameter of the planet is 500 kilometers. However, the object isn't completely round, measuring 476 km, 516 km and 550 km. Because of its shape, which is oblong the body isn't considered as an asteroid, but simply an asteroids.

Then, two years later in the year 1804, German astronomer Karl Ludwig Harding came across what's now the Juno asteroid. Juno having a mean size of 247 km.

It was in 1807 that Olbers found his planetoid within the Asteroid Belt, which lies located between Mars as well as Jupiter. Vesta has a median area of 525 kilometers.

From all of the minor planets that lie Between Mars and Jupiter of all the minor planets, the only one Ceres is thought to be an asteroid because of its massive size and near-spherical form. At present, thousands of planetoids exist. some of them located in between Mars and Jupiter however, others are in extremely elliptical orbits around the Sun.

It was in 1821 that French Astronomer Alexis Bouvard discovered some discordances regarding the location of Uranus in comparison to the parameters which had been calculated for it. He then proposed that there might be another planet in the vicinity of Uranus which was tugging at the same time on the earth Herschel was able to discover. British Astronomer John Couch Adams estimated the dimensions of the mysterious planet in 1843. Later, independently of Adams, French astronomer Urbain Le Verrier came up with his own parameters for the alleged planet. Le Verrier sent his estimate to an German scientist named Johann Gottfried Galle and asked to verify his estimate. Galle discovered the planet was just 1 degree off of the expected position. Le Verrier named the planet Neptune and it is identified as having a distance of 49,244 kilometers (3.86 times larger than Earth). The collaboration of Le Verrier along with Galle resulted in some of the most fascinating technological breakthroughs of the late 19th century. The team demonstrated how the theory (the mathematical foundation of the celestial mechanical system) could lead to discoveries (the space itself).

Le Verrier

Galle

In the 19th century, scientists and scholars thought about stars and planets of the universe and wondered whether any were alive and how they could one day go there to learn more. The early dreamers included Konstantin Tsiolkovsky, Jules Verne and Robert Goddard. Tsiolkovsky (1857-1935) was an Russian scientist who dreamed of stages for space stations and rocketry almost a century before they became a reality. Goddard (1882-1945) is a pioneering American scientist who was a pioneer in rocketry. He began exploring his concepts to launch rockets to the heavens getting closer and closer to the center of space. To his credit, Verne created a new category of fiction referred to as science fiction. his novels captivated the minds of many millions throughout the world. His novel of 1865, From the Earth to the Moon It was a novel that took readers to Earth's satellite prior to the time that airplanes were ever even allowed to visit.

It's clear that the inability to develop necessary technologies did not stop people from imagining their own. Russian astronaut Konstantin Tsiolkovsky envisioned humans living in space, and attempted to come up with ideas to allow it to be possible, like airlocks, steering thrusters as well as space stations.

In the 1930s and the 1920s, American inventor Robert H. Goddard in the 1920s and 1930s launched more than thirty rockets into the sky. Both his liquid fuel rocket as well as multi-stage rocket designs helped pave the way for space travel. While no of his rockets reached a height of more than 2 miles, his tests showed that his technology was a viable option.

Goddard

In the year 1930, Clyde W. Tombaugh discovered the dwarf planet Pluto and for over 70 years was referred to as the regular planet. Then, in 2006, the name was changed as scientists changed their definition of planet and came up with the definition of dwarf planet.

In addition, writers of science fiction in the late 20th century captivated youngsters with tales about interstellar space travel while the concept of flight was just beginning to take off. Television broadcasts featured shows such as Rocky Jones Space Ranger and Flash Gordon, and movies were beginning to show interstellar travel in a more realistic manner including for instance the 1955 MGM movie Forbidden Planet. It also featured animations from an artist from Disney artist who was loaned just for the purpose of making.

Then, towards the end of the 1950s America began to invest in rockets in the hopes of reaching space. And the technologies that led to America's success on the space frontier can be traced up to World War II, when Nazi Germany employed rockets in its attempt to destroy England. Following the final shots of World War II were fired and the rebuilding process for Germany and Europe started in the 1950s, Europe and Germany, Western Allies and the Soviet Union both tried to get access to the most renowned scientists from the Third Reich particularly those who were involved in the fields of rocketry, missile technology and research in aerospace. Naturally it was an extremely delicate matter due to the fact that some of these German scientists weren't just actively Nazis and had also helped to make to aid the Nazi army to inflict terror on the globe. However during the war's final period and into the post-war period, those who were Western Allies formed a clear image of the Soviet state. Although they were forced to join with the Soviets however, the West began to realize that Communist Russia as a aggressive totalitarian power which was a real danger to a free Europe.

The Western Allies and the Soviets were aware of Hitler's V-2 rocket programme which was the precursor to ballistic missiles as well as the race to space. Both recognized the importance of these technology and wanted to protect the benefits of these technologies in their own interests. While the Soviets considered expanding their reach following the "Great American War" and as the U.S. military came to recognize that their supposed allies from present would be the foes of the future those who had knowledge of V-2 rockets as well as various other Third Reich military technology programs were seen as key elements in the upcoming NATO to Warsaw Pact standoff.

This led to the America-led "Operation Paperclip" on the Western part and resulted German researchers putting their knowledge in the hands to both the U.S. and other NATO members. Operation Paperclip aimed not only to gain the benefits of

German technological advances to those in the United States but also to stop them from being accessed by the hostile Soviets in the manner that General Leslie Groves enunciated: "Heisenberg was among the top physicists in the world as of the moment of the German split, he was valued more than the ten divisions made up of Germans. Should he have fallen into Russian control, he could prove invaluable to them."

The Western method, though self-interested generally met with acceptance of the German scientist's part. However the Soviet solution to Paperclip, Operation Osoaviakhim, employed the implicit threat of torture, imprisonment and even death, one of the most common tools employed by Stalinist Russia in order to compel aid from German engineers and scientists following the outbreak of war. They paid rich dividends for the Soviet state, in terms the possibility of at least temporarily achieving technological parity with its Western adversaries.

to say that Operation Paperclip had a profound influence on the Cold War and American history is an overstatement. The most widely-known instance of Operation Paperclip's "success" can be found in Wernher von Braun. Von Braun was a former member of an SS branch. SS who was a participant in the Holocaust and would later be referred to for being"the "father of the rocket sciences" and captivate the world with his visions of winged Space stations and rockets in the basis for a "new" Manhattan Project, one which NASA was to ultimately take on. Alongside the ballistic missile weaponization which developed throughout through the Cold War, von Braun's skills were utilized for one of the oldest space missions in history. NASA needed to also develop rockets that were capable of launching the spacecraft in the Earth's orbit before sending it to the Moon.

The Soviets were unable to develop rockets capable of this task. However, due to von Braun, NASA got it just right with its Saturn V rocket, which is still the most powerful rocket launcher NASA has ever employed.

Von Braun

Following World War II, America formed The National Advisory Committee for Aeronautics (NACA) however, when the Russians launch Sputnik in the fall of 1957, the government of the United States considered the threat serious enough to approve the creation of National Aeronautics and Space Administration (NASA) in the year 1958. Then, a few years later the president Kennedy promised to place one man on the Moon at the close of this decade.

The very first stage of every space mission is achieving the orbit of Earth, which is a process that will require rockets with enough power to propel a spacecraft into orbit and then accelerate it to a speed of around 18,000 miles an hour, which is the rate required to reach the orbit of Earth. The precision, speed and technological sophistication required to reach the orbit safely is difficult to attain even in the present and even half a century ago. Likewise, the two most devastating disasters of NASA including the loss of Space Shuttle Challenger in 1986 as well as the loss of the Space Shuttle Columbia in 2003 were caused by problems which occurred during launch.

When the spacecraft enters orbit, it'll eventually fall back to Earth when it doesn't quit orbit. The velocity required for an object to exit orbit is referred to as escape velocity. In order to reach the escape velocity, and thus get out of Earth's orbit, the spacecraft has to travel close to 25,000 miles an hour.

When engineers were developing rockets that could be capable of bringing objects to speed up and escape its velocity during the 1960s, mathematics as well as scientists had to figure out what the different planets were in their respective orbits, as well as when they would be there. In the case of Mars, for instance, as Mars and Earth are orbiting different around each other around the Sun, Mars is in the same location with respect to Earth and the Sun just once in 780 days. This is called the synodic interval that is essential to calculate it accurately. Spacecrafts that travel through the Solar System will be orbiting the Sun through their entire journeys and it's not as easy as shooting the spacecraft straight line from where other planets are. To get it right involves extremely complicated mathematical calculations as well as if the project isn't completed on time it could result in delays of almost every synodic cycle.

While the Russians defeated the Americans with launching their satellite in space,

Americans accomplished the far more challenging task of bringing men to the Moon 20 July 1969, as well as the landing of a spacecraft by robotics on Mars in July 1976. In the year 1976, NASA had probes heading towards space in the Outer Solar System as part of the Pioneer programme.

The first missions were devoted to the development and testing of solid launchers. Numerous failures, some quite shocking caused engineers and scientists being discouraged in the early years. Finding solutions to those initial problems was not easy, but when they were solved the next missions could then be sent into space for various other tasks, such as explore the outer space around the Earth's atmosphere, study the Earth in space, journey towards the Moon to discover the structures of interplanetary space as well as the characteristics of the solar wind, and so much more.

In the 1950s, the time when NASA was created in the late 1950s in the late 1950s, scientists were unaware about the space's properties and the possibility of humans surviving in the atmosphere of Earth. It was difficult to know about other planets within the Solar System, and what was known had resulted from vague views obtained through earth-based telescopes as well as through the turbulent atmosphere on Earth.

Numerous initiatives NASA executed helped to determine the possibility of space exploration and travel for example:

Mercury Programme (1959-1963) Manned solo missions to sub-orbits and orbits.

Gemini program (1964-1966) two-man missions into space.

Apollo Programme (1966-1972) three-man mission to prepare for and target the Moon.

Pioneer program (1958-2003) Pioneer Program (1958-2003) - Multiple unmanned mission - Lunar probes Interplanetary satellite "weather" detect, the outer Solar System, and Venus probes.

Ranger Programme (1961-1965) Unmanned photography mission to the Moon.

Mariner program (1962-1977) Unmanned missions to study the inside of the Solar System - Mars, Venus and Mercury. The last set of missions were named Voyager 1 and 2, broadening the program's scope into the outward Solar System.

Surveyor Program (1966-1968) Surveyor Program (1966-1968) spacecrafts that were not manned to the Moon in order to show the possibility of landing gently on the moon's surface. closest neighbor in space.

Chapter 2: Rocket From Thor-Able

The launch took place on the 17th of August 1958, the main rocket (Thor) was unable to function just 77 seconds into the mission, and since the launch failed and the launch was not successful, the "Pioneer 1" designation was removed. Scientists believed that the Thor booster's malfunction was due to an impeller on the turbo pump that broke that would have stopped the oxygen flow and caused a rapid reduction in thrust. At that point, the whole spacecraft, including its launcher lost altitude as well as direction control and slid towards the opposite direction. The sudden change in direction probably ruptured the oxygen in liquid tank. This could have caught fire by the fuel, causing chaos.

The weight of the spacecraft's launch was only 83.8 pounds. The spacecraft was expected to take 2.6 days to reach the Moon. When it arrived, the in-board rocket motor made of solid propellant was fired, accelerating the spacecraft to ensure that it was able to fall into orbit approximately 18,000 miles over the moon's surface. If the mission was success, researchers could have had an understanding of what astronauts can expect within the lunar region from the threat of micrometeorites. Also, it would have given photographs of the Moon much closer than anything before which included the very first images of the Lunar Farside that had never been observed by humans. We would be waiting until the 7th of October of 1959, the day that Soviet spacecraft Luna 3 became the first spacecraft to capture images of the previously unknown part of the Moon.

Pioneer 0 was the first and the only Pioneer mission to be launched in the United States Air Force. Every the subsequent Pioneer missions were commissioned by the brand new National Aeronautics and Space Administration (NASA).

Pioneer 1 mission was also known as "Thor-Able 2." Pioneer 1 mission was also known as "Thor-Able 2." which was a second time it was designed to go around the Moon. It was an alternative mission that was a follow-up to the unsuccessful Pioneer 0.

The launch took place on the 11th of October 1958 The Pioneer 1 mission had a flawless liftoff. But the system for guidance malfunctioned and caused one incident following another, resulting in inadequate momentum for the spacecraft to exit the orbit of Earth. As a way to salvage the spacecraft, so it might be able to accomplish one important objective, NASA fired the onboard rocket motor. But its trajectory did not precisely align towards the higher Earth orbit required by it was aiming for in its own impromptu mission. The spacecraft instead hit a peak that was parabolic, which is the equivalent of 113,800 km (70,712 miles) before crashing back towards Earth. Pioneer 1 proved successful in observing the magnetic bands that surround the Earth. It also provided scientists with an understanding of "the Van Allen radiation belt, although technically speaking, Pioneer 1 "missed the Moon." In fact the spacecraft was not able to reach that point However, quick thinking helped the brand-new NASA agency to declare an insignificant victory on its initial Pioneer spacecraft mission.

An image from Pioneer 1 prior to launch

The probe's construction

Pioneer 2 was also called "Thor-Able 3" with the goal to circle the Moon and was the third effort to achieve the same thing that Pioneer 0 and Pioneer 1 did not succeed in doing. The launch took place on the 8th of November 1958, the lifting off of the second and first stage proved success, just like Pioneer 1, but like Pioneer 1, the third stage was not without its problems. It was the third time that the spacecraft reached only an elevation of 960 miles before returning to the atmosphere above Africa.

Pioneer 3 Pioneer 3 mission had a differing goal than the prior three missions. The spacecraft's goal was to travel over the Moon but not try to reach the orbit of the Moon. In the absence of the motor for its own rocket it was smaller and lighter than 12.9 pounds. This was just a tenth of lunar orbiters from prior three missions. If all went as planned, Pioneer 3 would have transformed into an artificial an asteroid.

The launch took place on the 6th of December 1958, the liftoff off of the launchpad went as per plans, but the launchpad's first stage rocket engine was stopped 3.7 seconds too soon. This is why the spacecraft could not reach sufficient speed to achieve the escape velocity. Other systems did not work as expected, however the ground control system was able to take advantage of opportunities despite difficulties. Pioneer 3 could achieve an altitude of 102 360 kilometers. This is about a third of the route to the Moon as well as on its journey back towards Earth, Pioneer 3 mission controllers were able to observe the spacecraft's measurements than the Van Allen radiation belt. The spacecraft was also able to try out a trigger mechanism that could enable a camera to be utilized on future missions.

A photo from The Pioneer 3 probe

Pioneer 4 was the only moon probe that was successful in the Pioneer program. This made it the initial American probe to escape the Earth's gravity and create an orbit of the Sun. Its goal was to examine the area around and around the Moon to detect radiation. The tiny photocell in the spacecraft was scheduled to trigger upon the time Pioneer 4 passed to within 30 km of the Moon however, the spacecraft's closest distance was 58.983 kilometers (36,650 miles) and was not enough for the sensor to activate.

It was launched on the 3rd of March 1959, the rocket engine were in perfect working order However, the second-stage burn did not stop at the right time, and gave the spacecraft just a bit more energy than what it required. However, the mission was successful and gave researchers with radiation readings throughout the Moon as well as beyond. Telemetry data transmissions were in operation throughout 82.5 hours during the mission, covering an area of 658,000 kilometers (409,000 miles).

Pioneer 4 spacecraft Pioneer 4 spacecraft achieved an orbit around the Sun by using these conditions:

Semi-majority axis: approximately 164,780,000 km.

Perihelion - 147,000,000 kilometers.

Aphelion - 169,000,000 kilometers.

Eccentricity = 0.07109.

Inclination - 1.5 degrees.

The period is 398.0 days (slightly longer than Earth's calendar).

A replica from Pioneer 4 probe. Pioneer 4 probe The Pioneer P-1 mission was also referred to as "Atlas-Able 4A" as well as "Pioneer W." Its purpose was to orbit the Moon using a TV camera as well as a magnetic sensor however, on the 24th of September 1959, as part of a prelaunch testing, the rocket engine was destroyed at the launchpad. The spacecraft was not yet installed onto the launch vehicle, and later was used for it's Pioneer P-3 mission.

This Pioneer P-3 mission was also named "Atlas-Able 4" "Atlas-Able 4B,"" as well as "Pioneer AX." If the mission had was successful then this Pioneer P-3 probe would have launched into orbit around Moon and could have utilized sensors that could have detected radioactivity, micrometeorites, electromagnetic fields, magnetic fields that has different frequencies, and also for determining a more precise moon's mass. It was equipped with its own motors to propel it through adjustments to its course, so this would have been the nation's first attempt to test capability of remote spacecraft to maneuver.

In the launch on November 26, 1959, all went well up to T+45 seconds. The tension inside the shroud of spacecraft significantly exceeded that of the air pressure at the altitude. It caused an explosive compression, causing a part of the payload shroud -- an element of fairing made of fiberglass that was ripped off of the spacecraft launcher. In the protective shield the support struts and spacecraft did not have enough aerodynamics to stand up to the raging turbulent. In the end, the third stage rocket as well as the payload was removed from the launch vehicle which effectively ended the mission.

It was the Pioneer P-30 mission was also known as "Atlas-Able 5A" as well as

"Pioneer The Y." Its purpose was to complete lunar orbit by using a variety of scientific instruments. It was in a back-up mission for the unsuccessful Pioneer P-3. The mission was launched on September 25, 1960, problems began with the failure of the fuel controls to turn off the first stage. This led to the vehicle to keep firing until fuel for the first stage exhaustion. The second stage was ignited correctly however thrust decreased to zero. The engineers later concluded that the flow of oxidizer had been unable to function properly, and the fuel was left with nothing which to ignite. Experts from the mission predicted that the spacecraft sank somewhere within the Indian Ocean.

It was the Pioneer P-31 mission was also named "Atlas-Able 5B,"" as well as "Pioneer Z." The goal of the mission was to reach moon orbit and conduct tests similar to those that were planned for the failed Pioneer P-3 and P30 missions. It was launched on December 15th, 1960, the liftoff went smoothly and, after sixty-six minutes into the operation the spacecraft suffered an abrupt drop in the gas pressure in the liquid oxygen tank. Spacecraft crashed into the Atlantic Ocean roughly 16 kilometers away from Cape Canaveral.

As with all the non-manned space probes, every was one Pioneer lunar spacecrafts was destined to fulfill an extremely specific purpose which could have brought specific facts about space research even though they failed technically however, some offered various kinds of success. In one sense, every mission was an examination of the technology used to launch vehicles which would allow future missions to space more reliable. The same way every failure was overcome by investigations which revealed little details which would result in better strategies for the launch of new spacecraft.

In the context that was created by the Cold War and the heating up of the Space Race, one of NASA's most difficult tasks was to market a space mission that seemed to go towards space. The Moon, a planet Moon and the Sun were easy targets to promote, while deep space appeared too far away both literally and metaphorically. In the end, NASA scientists realized they required to learn more about deep space, in order to comprehend the vast space around the Earth and the Moon.

Pioneer 5 was also known by the names "Pioneer P-2,"" "Thor-Able 4" or "Pioneer V" at this point the goal was to investigate the space that was in between Venus as well as Earth. The mission was certainly less glamorous than flying over the Moon or any of the planets. Initially, Pioneer 5 was supposed to fly over Venus however technical issues made it impossible to launch the mission in the Venus launch window, which was in November 1959.

Once the technical problems were solved in mid-June 1960 Venus began to move away so that it became impossible for Pioneer 5 to keep pace. An Venus mission had be awaiting the following launch window to ensure that Venus as well as Earth are in their correct, relative position.

It was launched on the 11th March 1960. Pioneer 5 became the most effective mission in the whole sequence of "Pioneer-Able" missions. The mission itself was fault-free, particularly when in comparison to other Thor-Able launches. The small issues with the second stage weren't enough to affect the mission's overall success.

The launch weight for the spacecraft of 43 kg (95 pounds). The Spherical probe (0.66 millimeters in diameter) there were the four "paddle-wheel" solar panels grew to take advantage of sunlight electricity. The vehicle used for the launch was one of the Thor DM-18 Able IV, departing from Cape Canaveral's with LC-17A.

The spacecraft landed in an elliptical an orbit that was heliocentric between Earth and Venus using the following orbital parameters

Aphelion Aphelion - 148,570,000 km (0.9931 Astronomical Units).

Perihelion - 105,630,000 kilometers (0.7061 AUs).

Eccentricity = 0.1689.

Inclination - 3.35 degrees.

Time Period 311.6 days (more than 50 days less than one year).

In comparison, they are the major axis semi-distances from Earth as well as Venus orbits:

Earth - 1459,023 (1.00000102 AUS).

Venus Venus - 108,208,000 (0.723332 the AUs).

It is evident that Pioneer 5 traveled from slightly further away from that of the Sun than Earth up to a spot that is farther from the sun than Venus.

Through Pioneer 5, scientists learned that interplanetary space actually does possess magnetic fields.

Pioneer 5 carried four important scientific instruments:

The solar wind sensor to determine the amount of solar radiation that include electrons having more than 13. MeV.

Magnetometers are used to gauge magnetic fields. This includes the magnetic boundary that forms part of Earth's field, as well as the strength the magnetic field that exists between the planets. The detection range ranged from 1 microgauss and up to 12 milligauss.

Cosmic ray detector for measuring the energy-rich particles of the distant supernovae (dying stars).

Micrometeorite detector for measuring the size and velocity of dust particles striking the spacecraft. It was the only one that did not return any useful information because of the saturation of data systems.

Solar panels are small for continuous data transmission. Larger panels would have taken up too heavy during the launch, which meant that the data transfer rate was limited every day. Overall, Pioneer 5 was able transmit information back to Earth through the 30th of April 60. The signal became weak as a result due to the distance that was increasing between spacecrafts and Earth as background noise became an important portion of the data received. At some point, the process of it became impossible to separate the noise from the data. was impossible. The next test by Jodrell Bank Observatory Jodrell Bank Observatory was able to pick up a signal of Pioneer 5 on June 26 in 1960. However, the signal was weak enough to be able to gather useful information. In the year 1960, Pioneer 5 was 36.2 million kilometers (22.5 million miles) distant from Earth which was a record-breaking distance in the era of.

While Pioneer 5 was not the famous media adolescent that other missions did, it nevertheless was an important milestone in the American space program. It accomplished nearly every goal it set with it was a fairly smooth launch. Its boost in morale was more valuable than its gold weight. This was all happening just one year prior to when the first cosmonaut stepped into space. And about a year prior to the first American astronaut Alan B. Shepard made it to space on May 5th, 1961.

An image from Pioneer 5 with test equipment

In 1962, when the president John F. Kennedy gave his famous "We Choose to Go to the Moon" speech on the 12th of September 1962, the world was changed.

The entire nation was asked to help the mission of getting an astronaut on the Moon at the close this decade. President Obama spoke to the public about the obstacles to be faced, as well as how difficult they could be.

"We embark on the new ocean, since there's new information to acquire, and new rights that can be gained and have to be earned and utilized to benefit every person. Space science, as any other technology does not have a morality of its own. The decision to use it as an instrument for positive or negative effects is up to the human race, and only when that the United States occupies a position that is preeminent can we determine whether the new ocean will become one of peace, or a brand new, terrifying theatre of conflict. This isn't to say that we ought to or should be unprotected from the violent usage of space the way that we do not have protection against hostile uses of the sea or on land. However, I am saying that space is a subject to be explored and studied without fueling the flames of war and without making the errors that mankind has made by expanding his reach across the planet of ours.

"There there isn't any strife or prejudices, there is no international conflict within space to date. Space's dangers are threatening for all of us. The moon's victory deserves the highest praise for all humanity, and it's chance to work in peace could never happen again. What is the reason, according to some that it is to the Moon? What is the reason we do we choose to pursue this goal? Some might even wonder, what is the reason for climbing the highest peak? What made you decide to just 35 years ago, did you to fly across through the Atlantic? What is the reason why Rice continue to play Texas?

"We decide to travel to the Moon! We have decided to travel to the Moon...We opt to travel to the Moon during this decade, and take on other activities and not just because they're straightforward, but because they're hard. as that is how we will be able to measure and organize the very best of our efforts and abilities, as the challenge we're willing to take on and are not willing to delay and that we are determined to succeed in, and other challenges, as well."

A photo of Kennedy speaking during his speech.

There was a split among those who disagreed with Kennedy. Certain people believed that such an enormous undertaking would be expensive, and offered too very little. Some thought that the Russians needed to be eliminated no matter what - and not solely because of a sense of pride in the country and pride, but also because the Soviet Union was a dangerous threat to peace in the world and the sight of Soviet spacecraft flying over every day for a long time seemed to be a bit frightening for nearly all American person who considered the possibility.

Although the president Kennedy didn't have the time to fulfill his goal the majority of those in government had a strong commitment to his mission to complete it. Maybe because Kennedy had been killed, some believed that to fall in the way of Kennedy's dream could be a way to desecrate the memories of a president that was murdered at the height of his career.

President Kennedy witnessed the success conclusion of the Mercury Program, which consisted of spaceflights manned by one astronaut inside a orbiting or sub-orbital spacecraft. But he didn't get to witness the inaugural Gemini Program manned flight, that took place a year and a half following the time of his death.

Although the Gemini Program would test and enhance the capability of astronauts in performing extravehicular tasks (EVAs) and docking spacecraft vital skills required in the coming Moon landing, the other robotic missions carried out a large portion of the research needed to prepare for the planned human-powered Moon landings. Pioneer was already part of these early endeavors, but it was now the other programs that would be performing the bulk of the work in studying the Moon.

Ranger could be taking close glances at the Moon and deliberately smashing its camera into the lunar surface to capture close-up images while descending. In the future, Surveyor would test the possibility of landing softly on the moon's surface. What we learned from this Surveyor Program was especially important in the future unmanned missions to Mars and further.

In the year 1960, NASA established their Mariner Program to generate flyby observations of the outer planets (Venus, Mercury, and Mars), Pioneer was prepared to assume the task of a completely different kind. Scientists were keen to learn more about what the distance between the planets looked like particularly after having achieved some results in the field with Pioneer 5 in 1960. In all the focus at physical bodies like moons, planets and even the Sun in general, the empty space was left almost under-explored. It was the Pioneer Program would take up the challenge of a new one.

Between 1965 and 1969 from 1965 until 1969, the Pioneer mission was launched every year in order to increase the understanding of what is now known as "space weather." It was only in 1969 that the Launch of Pioneer E ended in failure (because it was a failure for which it was never assigned an identifier).

A majority of the spacecraft taken out of gravity well of Earth and put into a heliocentric orbit which has an aphelion (closest proximity to Sun) that ranges from 0.75 AU and 1.0 A.U., as well as an Aphelion (farthest away from the Sun) that ranges from 0.99-1.2 AU.

Pioneer 6 included these mission details:

Launch Date: December 16 and 1965

Mass of Launch: 146 Kilograms

Launch Place: Cape Canaveral The LC-17A

Launch Vehicle Delta-E

Onboard Power: 79 Watts

Pioneer 6 was sent into the orbit of a circular orbit, that had an area of 0.8 an AU. This makes it the first mission that had the aphelion as well as perihelion which are virtually identical. Just over thirty years after its launch, the spacecraft's main travelling-wave tube (TWT) was shut down. In the summer of 1996, Pioneer 6 was given the task of restoring the TWT.

In the next year One Deep Space Tracking Station was able to record the Pioneer 6's movement. Additionally, various technological instruments were activated to verify the fact that they still worked and included that of the Chicago University's cosmic-ray detector, as well as MIT as well as ARC Plasma analyzers. In the evening of the 8th of December, 2000 Pioneer 6 was called in about 2 hours, requesting the successful download of telemetry.

Chapter 3: The Artistry Of Pioneer

The following instruments were aboard the Pioneer 6 spacecraft:

Celestial Mechanics

Cosmic-Ray Anisotropy

Cosmic-Ray Telescope

Electrostatic Analyzer

Relativity Investigation

Solar Wind Plasma Faraday Cup

Spectral Broadening

Superior Conjunction Faraday Rotation

Two-Frequency Beacon Receiver

Unaxial Fluxgate Magnetometer

Pioneer 7 was a mission that had the following mission information:

Date of Launch 17 August 1966

Launch Mass: 38 kg

Launch Place: Cape Canaveral The LC-17A

Launch Vehicle Delta-E

Onboard Power: 79 Watts

The mean orbital radius of Pioneer 7 is 1.1 AU, and on the 20th of March,1986, Pioneer 7 flew to just 12.3 million km of the largest celestial object that year Halley's Comet. From this safely location, the spacecraft investigated how sun's wind as well as the hydrogen tail of Halley's Comet. The 31st March of nine years after, a station from the Deep Space Network successfully tracked Pioneer 7 and discovered that the spacecraft as well as one science instrument were operational.

The following instruments were found on the Pioneer 7 spacecraft:

Celestial Mechanics

Cosmic-Ray Anisotropy

Cosmic-Ray Telescope

Electrostatic Analyzer

Magnetometer with a single axis Magnetometer

Solar Wind Plasma Faraday Cup

Superior Conjunction Faraday Rotation

Two-Frequency Beacon Receiver

Pioneer 8 had the following information about the mission:

The Launch Date is 13 December,1967

Mass of Launch: 146 kg

Launch Place: Cape Canaveral The LC-17A

Launch Vehicle Delta-E

Power onboard: 79 Watts

The spacecraft placed to a high-altitude orbit, with a an orbital mean of 1.1 A.U. The 22nd August of the year 1996 Pioneer 8 was ordered to change its operation to its TWT backup (traveling-wave tube). Telemetry was purchased in 1996, and one of the instruments used for research was re-introduced after more than 30 years of space.

The following instruments were aboard the Pioneer 8 spacecraft:

Celestial Mechanics

Cosmic Dust Detector

Cosmic Ray Gradient Detector

Cosmic-Ray Anisotropy

Electrostatic Analyzer

Plasma Wave Detector

Magnetometer Single-Axis Magnetometer

Two-Frequency Beacon Receiver

Pioneer 9 included the following mission information:

Date of Launch 8 November 1968

Start Mass: 147 Kilograms

Launch Location: Cape Canaveral, The LC-17A

Launch Vehicle Delta-E

Onboard Power: 79 Watts

The spacecraft sent to a geocentric orbit, with an average distance to solar rays of 0.8 AAU. Pioneer 9's last contact successfully made with Pioneer 9 occurred in 1983. A second attempt was attempted in 1987, however it did not succeed.

The following instruments were aboard the Pioneer 9 spacecraft:

Celestial Mechanics

Cosmic Dust Detector

Cosmic Ray Gradient Detector

Cosmic-Ray Anisotropy

Electronic Field Detector

Solar Plasma Detector

Triaxial Magnetometer

Two-Frequency Beacon Receiver

The four spacecraft collect near-real time data about solar storms but they also provided satellite data regarding solar storms, which were being monitored by observers on the ground and instruments aboard listening balloons. The beginning of August in 1972, Pioneer 9 was at the perfect time and place to record information from its "front row" position near the largest and strongest solar storm that has ever been observed. Through all that study, scientists now have greater understanding of the extremes that astronauts may encounter from solar flares as well as coronal mass expulsions.

In the direction of out to Outer Solar System

In the process of preparing engineers and scientists for spacecrafts for the first time to conduct a mission, they relied on their intuition. They were not aware of the circumstances that the spacecraft might encounter during its journey which included the quantity of radiation and temperature extremes as well as the cosmic ray explosions as well as the intensity of magnetic fields that the probe had to contend with. Scientists were required to form some assumptions and engineers needed to adhere to the requirements based on those assumptions, in order to ensure the safety of the instruments of the spacecraft.

In the beginning the entire system had to endure the enormous heat and vibrations during the space launch. Everything had to be able to endure the extreme cold and heat in space, where the Sun may be shining at all the time but the dark space of space was frozen in the following.

Spacecraft telemetry has always integral to every spacecraft mission, no matter if it's a successful or not. Sensors onboard can help engineers and scientists understand the reason why missions fail and help develop a better spacecraft, or launch vehicle to carry out the future. Pioneer 10 and Pioneer 11 weren't any different. However, what they discovered when they entered the sub-system of Jupiter amazed them more than they ever could have.

In the early days of space travel was just beginning to become a thing the scientists as well as NASA administration were making plans for their biggest and most adventurous mission yet in order to explore beyond the internal Solar System of terrestrial worlds (Mercury, Venus, Earth and Mars, the Moon as well as Mars) as well as past the asteroid belt and finally, into the Outer Solar System of gas giants and giant ice planets. Pioneer 10 and Pioneer 11 were the very first two spacecraft designed to attain the speed of escape from solar radiation, which means that these spacecraft would to leave this Solar System at some date in the near future.

The early 1960s were when an early 1960s Jet Propulsion Laboratory aerospace engineer known as Gary Flandro initiated an idea to create an idea for a Grand Tour of the Outer Solar System. Because Earth and Jupiter move constantly and traveling at different speeds there are only certain periods where a space launch is accomplished with only a minimal requirements for fuel. Other times, it would need massive amounts of fuel in order to travel much greater distances to reach Jupiter as well as Flandro observed the alignment of planets in the 1970s could enable multiple slingshots and gravity-based aids, so that every one of the outer planets can be explored.

Two Voyager spacecraft benefited from the alignment. However, NASA was keen to be prepared for the chance to be lucky through sending two previous probes to the dim, cold realm of giant planets. So they Pioneer probes will be the first probes to explore the asteroid belt, and identify what hazards are present that might affect any future missions. While these two Pioneer spacecraft wouldn't complete the Grand Tour, their work could be vital in preparing the Voyager spacecraft for the tough challenges that lie ahead. The instruments used by scientists aboard the spacecraft were designed to capture images of Jupiter and the moons it has as well as make several observations across a variety of wavelengths of ultraviolet, infrared and the polarized light. The instruments were designed to find magnetic fields, charged particles cosmic rays, plasma as well as zodiacal light. The research would consist on the planet's atmosphere as well as its weight, and also the masses of moons.

NASA's Ames Research Center was given the task of managing the project. For the sake of speeding up the speedy development of the spacecraft Ames overcame the standard procedure to submit a Request to Proposals and opted for TRW Inc. for their impressive experience to that moment. A humorous exchange occurred, one TRW engineer stated regarding the Pioneer spacecraft "This spacecraft will be guaranteed to last for two years in interplanetary space flight. If any component fails in the warranty time, simply take the spacecraft back to us and we'll fix it for no cost." Naturally the declaration was completely joking as no one would have been capable of relocating the spacecraft to take them to repair that would have been much more expensive as the entire mission.

Although the initial plan of Grand Tour Grand Tour did not include the use of slingshot effects to leave from the Solar System, such an concept was included in the Pioneer mission prior to the launch. If everything goes as planned, the project managers could modify the trajectory of a mission in order to complete the mission.

Pioneer 10 and Pioneer 11 featured similar instrumentation packages to study the properties of space during its travels and also in the close vicinity of Jupiter (and for Pioneer 11, in the area also around the planet Saturn). One of the instruments that was on Pioneer 11 was not also present on Pioneer 10: the Triaxial Fluxgate Magnetometer.

Helium Vector Magnetometer (HVM) The HVM studied the intricate design of the magnetic field that runs across the planet and plotted the intensity of the magnetic field on Jupiter at different locations and also made measurements of the Jovian magnetic field to determine how the solar wind acted on Jupiter, the biggest planet in the Solar System.

Principal investigator: Edward Smith / JPL.

Access to data at the PDS/PPI's data catalog and NSSDC archive of data.

The HVM

"Quadrispherical Plasma Analyzer - Designed to an opening in the dish antenna of the spacecraft for the purpose of capturing solar wind particles, which originated from the sun.

The primary investigator is Aaron Barnes / NASA Ames Research Center (archived site).

Access to data at the PDS/PPI's data catalog and NSSDC archive of data.

It is the Quadrispherical Plasma Analyzer

The Charged Particle Instrument (CPI) was used to measure the power of cosmic rays on their path across the Solar System.

The primary investigator is John Simpson / University of Chicago.

- Data available at NSSDC data archive.

Cosmic Ray Telescope (CRT) The telescope has collected information about the many kinds of cosmic ray particles as well as the spectrum of their energies.

Principal investigator: Frank B. McDonald / NASA Goddard Space Flight Center.

The data is available on both PDS/PPI's data catalog and NSSDC archives of data.

Geiger Tube Telescope (GTT) Protons and electrons were measured across the entire spacecraft's route as well as the distances distance across and through the Jovian as well as Saturnian (Pioneer 11, only) radiation belts. The measurements included angular distributions along with energy spectrums and intensities.

Principal investigator: James A. Van Allen and the University of Iowa.

The data is available through three different sources: the PDS/PPI Data catalogue, NSSDC archives for data, NSSDC Jupiter archive of data.

Trapped Radiation Detector (TRD) It included three different instruments with a minimum-ionizing detector that was a solid-state diode, which measured protons that range from 50-350 MeV as well as the smallest ionizing particle (<3 MeV), an electron scatter detector to detect electrons with energies ranging between 100 and 400 keV. It also had an non-focused Cherenkov counter that analyzed the radiation that was emitted in a specific direction, as particles passed through it, recording electrons with energy that ranged between 0.5 -12 MeV.

The primary investigator is R. Fillius / University of California San Diego.

The data is available on NSSDC daily data archive. NSSDC Saturn data archive.

Meteoroid Detectors, which measured the impact of penetrating meteoroids. The device was comprised of 12 panels that were positioned on the back of the dish's main antenna (pressurized cells as detectors).

Principal Investigator: William Kinard / NASA Langley Research Center.

Access data from NSSDC archives of data.

Asteroid or Meteoroid Detector (AMD) tracks particles of various sizes ranging from close dust motes, to distant massive asteroids.

Investigator: Robert Soberman / General Electric Company.

Access data from NSSDC archives of data.

Ultraviolet Photometer to determine the quantity of hydrogen and Helium within space as well as on two planets, Jupiter as well as Saturn. It was possible because each gas has emission spectra within the UV band.

Principal Investigator: Darrell Judge / University of Southern California.

Access to data can be found in the PDS/PPI data catalog as well as NSSDC archives of data.

Imaging Photopolarimeter (IPP) Imaging Photopolarimeter (IPP) The program took pictures of the Earth, using an extremely small telescope many times on the surface of the object while the spacecraft rotated. Each time, it only captured 0.03 degrees of the arc. Two series of complete images were taken using an orange filter, the other that had the blue filter. Through the many strips that were in the series, an entire image was created to give an all-color image of our planet's landscape.

Principal investigator: Tom Gehrels / University of Arizona.

Access data from NSSDC Data archive list.

The Imaging Photopolarimeter

Infrared radiometer, which provides information on cloud temperatures and the excess heat emissions of the planets. Jupiter is an example, and has been emitting warmth from the initial compressing of its beginning 4.6 billion years long ago.

Primary researcher: Andrew Ingersoll / California Institute of Technology.

Chapter 4: The Infrared Radiometer

Triaxial Fluxgate Magnetometer (Pioneer 11 only) The instrument measured the magnetic fields of Jupiter and Saturn.

- Primary investigator: Mario Acuna / NASA Goddard Space Flight Center.

The data is available on the NSSDC archives of data.

Pioneer 10 had the following information about the mission:

Date of Launch February 2, 1972.

Launch Vehicle Atlas SLV-3C Centaur Star-37E.

Launch Location: Cape Canaveral The LC-36A.

Launch Mass: 258.8 kilograms (571 pounds).

Power onboard: 155 Watts (at start).

Last telemetry received: the 27th of April, 2002.

"Last Signal: 23rd January 2003.

The third component of the launcher (TE364-4) is which is a solid fuel booster was developed specifically to support the Pioneer outer planets spacecraft specifically. In addition to delivering the necessary 15,000 pounds thrust, it also produced an orbit for the spacecraft that was able in achieving gyroscopic stability throughout the duration of its mission. The initial spinning speed was 30 minutes of rotation (rpm) However, around 20 minutes after it's launch Pioneer 10 extended its 3 booms and then slowed its speed to 4.8 per minute. The standard speed of rotation throughout the mission.

Once the launch vehicle used up its fuel after 17 minutes in flight it had reached speeds of 51,682 km in an hour (32,114 miles per hour). This was significantly higher than the 40,270 kilometer per hour required escape speed to be achieved. Not long afterwards ground-based mission control made contact with Pioneer 10's high-gain antenna instructing it to activate a number of Pioneer 10's sensors for testing as it travelled through the radiation belts that surround Earth.

The launch photo

Pioneer 10 officially reached interplanetary space approximately 90 minutes after launch. The spacecraft had exited gravity well on Earth (the place where the Earth's hold on the spacecraft wasn't as powerful as it was for its own Sun). After 11 hours of the launch, Pioneer 10 had passed the moon's orbit. Moon as well, and at this point, it was the fastest thing that humanity could ever have created.

Within 48 hours of the the launch, mission experts began turning on the rest of the scientific instruments, beginning at the beginning with CRT. On the 10th day of launch it was the time that all scientific instruments were activated.

In June of 1972, Pioneer 10 crossed the orbit of Mars and increased its distance to the Sun every day. On July 15th, 1972, Pioneer 10 had officially reached the asteroid belt. This was being the first human spacecraft complete this achievement. Additionally, around this time the spacecraft began the Jupiter observation phase. However, it was still over a year to go before hitting the massive planet.

Between March and the close September 1972, Pioneer 10 completed three courses of adjustments to make sure it could get to its target. In this time, the Onboard equipment went through a variety of checks, which included the measuring of cosmic rays and solar wind and magnetic fields. They also had cameras that observed the light from the zodiacal sphere and, the primary target of the mission, Jupiter. One notable error that involved a camera aimed at the star with the highest magnitude, Canopus. It was designed to ensure the perfect alignment of every system to ensure that the instruments used for research would be pointing in the proper direction. Due to the Canopus sensor malfunction, engineers suggested that the spacecraft be equipped with the two sensors on solar panels instead.

Pioneer 10's photo of Jupiter

Unexpectedly, Pioneer 10 was able to achieve a variety of milestones. One of them was finding elemental Helium along with high-energy sodium as well as aluminum ions within the medium between planets. Pioneer 10 was the first emissary of Earth to be able to penetrate the space-based asteroid belt on the 15th of July 1972. It traveled through the sphere of a tiny dwarf planet (Ceres) and many planetoids and many meteoroids that orbit those orbits Jupiter as well as Mars. Project planners for the Pioneer Outer Planets Project were confident that Pioneer 10 to remain safe all the way across the asteroid belt due to contrary to the depictions in films the asteroid belt appears exceptionally thin. From all of the known asteroids discovered over the past 2 centuries, only a handful could be nearer than 8 million kilometers (5 millions miles) in proximity to Pioneer 10.

In the beginning in August of 1972 while Pioneer 10 was at a distance of 2.2 Aura from the Sun within the belt of asteroid It recorded a unique sunshock wave.

As for the asteroid belt scientists' instruments showed that in comparison to the surrounding area of Earth the belt was deficient in dust particles smaller than a millimeter (mm). The density of dust particles between 10 and 100 mm was relatively stable across Earth up to the extreme area in the belt of asteroid. There was only one surprise in the diameter of particles of 100 mm up from 1.0 millimeters (mm). The belt had a threefold growth in particles with the size of this, however fortunately particle sizes greater than millimeter did not appear anywhere in the belt. Pioneer 10 emerged on the edge of the belt in February 15th, 1973.

When the spacecraft cleared its belt NASA tried to make significant changes to its course. The new flight path will permit Pioneer 10 to use Jupiter as a gravitational boost as well as a slingshot that will take the spacecraft away from its Solar System entirely. This drastic change to its trajectory proved an intricate maneuver was able to be done remotely, and it became the model for any the future space missions that may need to perform a similar maneuver.

Pioneer 11 had the following information about the mission:

Date of Launch June 6, 1973.

Launch Vehicle: Atlas SLV-3D centaur-D1A Star-37E.

Launch Location Cape Canaveral Cape Canaveral LC-36B.

Launch Mass: 259 kilograms (571 pounds).

Inboard Power: 155 W (at start).

Last Contact 30 September 1995.

Pioneer 11 was the Pioneer 11 launch vehicle would do all the work, launching the spacecraft directly towards Jupiter with no gravitational assistance during the journey. Spacecraft Pioneer 11 began its course adjustments on the 11th of April 1973. This was eight months before Pioneer 10 was to reach Jupiter and twenty months prior to its arrival near Jupiter, the planet with the most powerful crown.

Around April 19th, 1974, Pioneer 11 had officially finished its journey through the Asteroid Belt. A few weeks later, in May, following the fact that its sibling Pioneer 10 Pioneer 10 had already completed its Jupiter mission, Pioneer 11 was given an entirely new direction north-south that could allow it to come the asteroid Saturn in 1979. The maneuver used up nearly eight tonnes (17 pounds) of propellant. The thrust added took approximately 42 minutes and 36 secs which increased the spacecraft's speed at 230 km/h.

The 3rd November, 1974, Pioneer 11 officially began the Jupiter Phase of observation. The next day Pioneer 11 completed the final correction to its mid-course.

Illustration of Pioneer 11 in space

Pioneer 10 began testing its imaging technology on the 6th of November 7th, 1973. To the east, Jupiter lay waiting for the arrival of its satellite, and there were 25 million km (16 millions miles) left to travel. In the meantime, the Deep Space Network back on Earth was able to successfully receive the images stream.

In the wake of this accomplishment, NASA mission specialists uploaded 16,000 commands into the computer of the spacecraft, instructions that will control spacecraft's operation as it moved through the Jovian subsystem for the next two months.

On the 8th of November, Pioneer 10 passed the orbit one of Jupiter's largest moons. It was Sinope. On November 16, Pioneer 10 was at the bow shocks of the Jovian magnetosphere. Scientists can tell this from the abrupt reduction in the solar wind speed between 451 km/s (280 miles per second) up to 225 kilometers per second (140 miles per second). In the following 24 days, Pioneer 10 reached the magnetopause and found Jupiter's magnetic field opposite to that of Earth (with the south magnetic pole being pointed in the similar direction to Earth's north magnetopole).

As the spacecraft grew closer to Jupiter it's spinning camera could scan parts of images in blue and red hues employing the imaging photopolarimeter. In addition to the two hues there was a third green color was synthesized, which allowed the full-color

RGB (red-green-blue) picture to be created. 12 of these images were acquired by the Deep Space Network back on Earth. It happened within a mere week of the center of its mission.

On November 29, 1973, Pioneer 10 had gone beyond the orbits of all of Jupiter's moons with the outermost orbits. All systems were operating perfectly.

Just a bit more than a full day until the closest encounter of Jupiter, the images that came through Pioneer 10 easily exceeded the highest quality images created by Earth at that time. Television news coverage reported the event live showing the images live when they arrived back in Earth. The Pioneer show later earned the privilege of winning the highly coveted Emmy award for its broadcast to the public at large. While the speed of the spacecraft produced distortions to the images, these could be easy to correct in the future by an image-correction algorithm. In the course of the

Jovian sub-system interaction, Pioneer 10 transmitted more than 500 photos of the gas giant as well as its moons.

3 December 1973, Pioneer 10 began its meeting with The Jovian sub-system. The spacecraft was launched at 12:26. passed by the most distant Galilean moon called Callisto (4,821 km in size and 16.69 day orbit around Jupiter). Nearest distance was 1,392,300 km, close to 500,000 kilometers more than the moon's distance to the planet it shares with.

The time was 13:56. Pioneer 10 flew by the following Galilean moon Ganymede (5,262 kilometers, 7.15 day orbit). Nearest approach was 446,250 km, which is roughly 600,000 km closer to the distance of Ganymede to Jupiter.

Pioneer 10's image of Ganymede

The spacecraft made its closest approach Europa, the 3rd Galilean moon Europa (3,122 kilometers, 3.55 day orbit). From an area of 321,000 km, Pioneer 10 was roughly just half the distance that separates Europa from Jupiter.

At 22:56, nearly 10 hours following the Callisto meeting, Pioneer 10 came to close to 357,000 km of the farthest Galilean moon Io (3,643 kilometers, 1.77 day orbit). It was just 60,000 km further from Io than Io is closer to Jupiter.

Scientists wanted to know more about the magnetic field of Jupiter and the radiation ionized around the magnetic equator of Jupiter, the huge planet. This is why Pioneer 10's trajectory took it across the equator of Jupiter's field. On Earth the scientists were shocked to discover that the peak electron radiation flux was almost 10,000 times higher than the maximum flux in the area of Earth. Engineers weren't prepared for this kind of radiation and they were also unable to have any method of determining this prior to the event.

The two Voyager spacecraft were to take the data into consideration in order in order to shield its equipment from damage however, in the meantime it was the Pioneer spacecraft was required to take the punishment as well as it could. Through its journey between the Jovian sub-system to another, Pioneer 10 suffered multiple instances of incorrect commands produced by the computer onboard system that was reacting to radiation. A majority of the mistakes were rectified by subsequent commands which were incorporated to deal with unexpected events. However, one image of Io along with a couple of good pictures of the globe could not be recovered due to the issues. The end result was that Pioneer 10 was successful in making fairly clear photos of Europa as well as Ganymede.

A beautiful rendition of Pioneer 10 near Jupiter

In the early hours of 2,26 a.m. on the 4th of December in 1973, Pioneer 10 came to just 200,000 kilometers from Jupiter making it the first spacecraft originating from Earth ever to witness Jupiter, the planet with the most powerful crown at close range. In this closest distance it had increased its acceleration as a result of its "fall" towards the massive planet, speeding up by 132,000 kilometers an hour. Imaging instruments captured clear images of the Great Red Spot and the planet's terminator. Ten minutes later, Pioneer 10 crossed Jupiter's the equatorial plane.

at 2:41:45 Pioneer 10 was perfectly positioned to occult with Io in view from Earth. The innermost Galilean moon is blocking the view of Earth's Pioneer 10, an opportunity which gave researchers from Earth an opportunity to check the moon's atmosphere. at 2:43:16, Pioneer 10 exited its Io Occultation and was once more making contact line-of-sight Earth as well.

By the time it was the scientists could discover that the ionosphere of Io is approximately 700 km (430 miles) in altitude when it is on the sunny side. Researchers also observed that electron density was near 60,000 cubic centimeters on the bright side, while density decreased to 9,000 on the dark side.

The scientists were pleasantly surprised by one of the discoveries that was the nature of Io's orbit about Jupiter. The moon is in a cloud of hydrogen which extends to the west by about 805,000 km (500,000 miles) in height and the width of around 402,000 km (250,000 miles). Researchers also discovered evidence of that there is a similar cloud in the Europa orbit. Europa.

The spacecraft entered its occultation of Jupiter which provided a large quantity of details about Jupiter's gas giant's atmospheric layer as radio signals moved across the upper layers in the clouds above, in their return to Earth. Then the spacecraft entered Jupiter's shadow. This blocked its view of the Sun. In the 4:05 minute mark, Pioneer 10 exited its orbit around Jupiter. At 4:47:21, the Pioneer 10 exited Jupiter's shadow.

Occultation in Jupiter enabled scientists to study the structure of Jupiter's temperature in its outer atmosphere. They found a temperature shift between the pressure levels 100 millibars and 10 millibars. For the 10 millibars levels temperatures ranged between frigid range of -133 to -113degC (-207 to 171degF). When you get to 100 millibars levels temperatures were more frigid, with temperatures ranging between -183 and 163 degC (-297.4 between -297.4 and -261.4degF). Based on the temperature data, researchers were able to construct an image of infrared radiation on Jupiter. With the help of this map they were able prove the fact that Jupiter releases more than the Sun and the residual heat resulting due to the compression that began its creation over 4 billion years prior.

While Pioneer 10 moved away from Jupiter and took photos of the dark side, which showed the crescent moon, the perfect ending for this spacecraft's journey to the most powerful planet.

In January of 1974 Pioneer 10 was officially finished in its Jupiter mission. But between November 30 and December that year, Pioneer 11 passed through the Jovian subsystem.

The 2nd of December, 1974, Pioneer 11 began its meeting with its Jovian sub-system. The spacecraft, at 08:21, passed by the farthest Galilean moon Callisto (4,821 km in size, 16.69 Day orbits around Jupiter). Closest approach was 786,500 kilometers. That's approximately half of the distance traveled by the closest orbit of Pioneer 10.

The time was 22:09 when Pioneer 11 flew by the following Galilean moon Ganymede

(5,262 km; 7.15 Day orbit). Nearest approach was 692,300 kilometers. This is nearly twice the distance that Pioneer 10 flew by on its closest approaching.

The Jupiter sub-system's fly-through continued until December 3rd, 1974. At 3:01, the spacecraft completed it's Io fly-by (3,643 kilometers, 1.77 day orbit) which was the most inner Galilean moon. The closest encounter was at the distance of 314,000 km which is five times that from the nearest approach by Pioneer 10.

Pioneer 11's image of Io At 4:45, Pioneer 11 flew past Europa (3,122 km, 3.55 day orbit). Nearest approach to the moon was 586,700 kilometers. This is almost twice as far as Pioneer 10 from that moon when it was closest to the moon.

At 5:21 pm, the spacecraft slid into Jupiter's shadow, stopping it from preventing Sun from shining upon the craft.

At 5:01:01 Pioneer 11 was occulted by Jupiter when seen from Earth which blocked every radio connection between the spacecraft as well as mission control.

In the 5th minute of time, Pioneer 11 had its closest encounter with Jupiter close to 42,828 km (26,612 miles) of the Jovian cloud the tops. After the flight, Pioneer 11 was able to take extremely detailed images of the Great Red Spot, plus clear pictures of the regions with polarity. Pioneer 11 could also help identify the precise size of Callisto.

At 5:33:52, spacecraft left the shadow of Jupiter and was enjoying the glow that emanated from the Sun.

At 5:43:03 the probe was able to leave at 5:43:03, the probe left Jupiter occultation. It was capable of transmitting its telemetry back to Earth.

In the 22nd minute, Pioneer 11 flew to close to 127,500 km of Jupiter's moon Amalthea.

In January of 1975, Pioneer 11 was officially completed in the Jupiter part of its mission.

In utilizing the gravitational pull from Jupiter to increase the speed of its orbit and alter its trajectory, the gravity aid altered its trajectory so that the probe could be pointed precisely towards Saturn.

Three and a half years following, on the 16th of April in 1975, technicians from the project shut off the micrometer detection.

A little over two years later Voyager 1 was launched from Earth with a stronger rocket on the 5th of September 1977. In March of 1979, Voyager 1 completed its own Jupiter flyby. About four months following, Voyager 2 completed its Jupiter flyby. Both Grand Tour spacecraft were then heading towards Saturn which was scheduled to enter Saturn's ringed planet by late 1980 or late in 1981 respectively.

Pioneer 11's photos of Jupiter

Because both Voyager spacecraft were successfully in Jupiter and were near Saturn, Pioneer project planners took the decision to try something that was extremely dangerous with their spacecraft. They decided to explore the spaces within the structure of the ring in order to assess how risky they could pose to Voyager spacecraft to come. Nearest approach to Saturn clouds was around 21,000 km (13,000 miles). It is the center of the "C" rings of Saturn's beautiful ring however, at the closest point it was not close to the plane of the circle. It travelled beneath the stomach of Saturn and then travelled through the ring's plane ascending mode before arriving at the nearest to the planet, and then crossing over the ring's ring plane in an ascending mode. It was farther away from Earth, yet still quite near to the edges of the circle. The mission was one of space-based "chicken," and potentially very expensive. However, considering two larger and better equipped probes following it and the risk of Pioneer

11 seemed worth it. Pioneer 11 would be going in a similar direction as those that are planned for coming Voyager trips, and should Pioneer 11's Pioneer mission proved dangerous it would be Voyager 2 would be able to take a safer route. Voyager 2 spacecraft will follow a more secure route (and therefore miss having the chance to see Uranus as well as Neptune).

Every exploration comes with dangers. In the case of Pioneer 11, scientists didn't realize just how dangerous it was to launch an asteroid-laden spacecraft into the belt before they sent Pioneer 10 there. In the future, Pioneer 11 would become an "pioneer" to the fullest meaning of the term and would set the course to the next two Voyager probes to come one year and two years after.

On the 31st day of July 1979, Pioneer 11 began its Saturn observation phase. It was a time to take photos and testing the region that would eventually become the Saturnian subsystem.

On the 29th of August 1979, Pioneer 11 officially entered the Saturnian subsystem.

At 6:06:10 the probe passed by Saturn's moon Iapetus (1,470 kilometers in diameter, 3,608,820 kilometers of circular radius). The closest point of contact was 1,032,535 km.

The same day, at 11:53:33, Pioneer 11 flew by Phoebe (213 kilometers in diameter 12,869,700 radius orbital) If you're able to be truly define it as an "flyby." In the nearest distance was 13,713,574 km, roughly one third in the space from Earth and Venus when they were at the closest point. It was akin to running along the boulevards of Los Angeles, waving to people who was in New York City. We used the word "flyby" in this instance is a bit of a stretch.

The next day, September 31, 1979 the probe made another encounter. The time was 12:32:33. Pioneer 11 flew as close to reach Hyperion (270 km diameter and 1,481,010 kilometers of the orbital radius). The distance was 666.153 kilometers.

The next day, on the 1st of September 1979 Pioneer 11 entered the heart of the Saturnian subsystem.

At 14:26:56, the probe travelled across the ring plane while in descend direction.

at 14:50, Pioneer 11 flew by the new moon, which was previously not discovered. It was named "Epimetheus" (116 km in diameter and 151,422 km circular radius). Its closest point was only distance of 6,676 kilometers. It was close to collision which may stop Pioneer 11 at a cold. In a strange twist, Epimetheus was discovered within the same orbit as Janus but in a different orbit at a later date.

At 15:06:32 the spacecraft flew past Atlas (30.2 kilometers in diameter and 137,670 kilometers of the orbital diameter). The closest point of approach to the spacecraft was at 45,960 km.

The probe was at 15:59:30 when it took off from Dione (1,122.8 km radius, or 377,396 km orbital diameter). Closest distance was 291,556 kilometers.

At 16:26, Pioneer 11 flew by Mimas (396.4 kilometers diameter, 185.404 kilometers of orbital radius). The closest point of contact was 104,263 km.

At 16:29:34 the spacecraft was able to make the fastest approach to Saturn close to 20591 km of the tops of the clouds, which is far from the surface of the ring.

In 1635 Pioneer 11 started its radio blackout. It was being blocked by the body Saturn. The artificial occultation placed the planet in between Earth and the envoy of its.

At 16:35:57, the spacecraft passed through Saturn's shadow, putting the planet's body between the spacecraft and Sun.

At 16:51:11 Pioneer 11 flew by Janus (179 kilometers in diameter, 151,472 km orbital radius). Nearest approach was 228.988 kilometers.

At 17:53:32, spacecraft restored line-of sight connection to Earth without causing its false Saturn occultation.

At 17:54:47 the probe was able to leave Saturn's shadow.

At 18:21:59 Pioneer 11 made its ascent rings plane crossing. The distance between the ascending the ring crossing (approach) and the closest distance to Earth was 2 hours 2 min. 38sec. The duration between the nearest Saturn approach as well as ascending rings cross-section (departure) was 1 hour 52 minutes. 25sec. This means that the route was an essentially symmetrical path through the solar system and its rings. The departure side was the departing side just a bit further away from the rings.

The time was 18:25.34. Pioneer 11 flew by Tethys (1,062 kilometers in diameter 29,619 kilometers in orbital radius). Closest to the ground was 329,197 km.

At 18:30:14, spacecraft was flying by Enceladus (504.2 kilometers diameter, 237.950 kilometers in orbital radius). The closest point of contact was 222.027 kilometers.

The probe was at 20:04:13 when it passed through Calypso (21.4 kilometers in diameter and 294,619 km orbital radius). Nearest approach was the distance of 109,916 km.

At 22:15:27 Pioneer 11 flew by Rhea (1,530 kilometers in diameter, 527,108 kilometers of orbital circle). Closest flight was about 345,303 kilometers.

A full day later the 2nd of September 1979 the craft completed the closest approach to Titan, Saturn's biggest moon. Titan (5,149 kilometers in diameter 1221,930 kilometers of circular radius). It made its closest approach at 18:00, which is 362,962 km. Titan is, the dimensions of a tiny planet it is certainly cold enough for life to survive, but it has been discovered later to be home to a surface air pressure that was significantly higher than the one observed on Earth.

Pioneer 11's image of Titan

As it passed through the Saturnian subsystem, in addition to being astonished by its interaction to Epimetheus, Pioneer 11 also found a small moon as well as another ring that was not previously discovered.

On the 5th of October 1979, Pioneer 11 was officially finished in the completion of its Saturn mission.

Pioneer 11's photos of Saturn

The end result was that Pioneer 10 and Pioneer 11 were not impacted by the moons of small size or asteroids. In addition, thanks to their gravitational enhancements They were able reach escape speeds away from the Sun and also from the Solar System.

While Pioneer 10 was not headed towards Saturn however, it was able to reach the orbit of Saturn in the year 1976. It then reached its orbital radius of Uranus in 1979.

On the 25th of April of 1983, the probe was able to pass through its orbit around Pluto and Pluto, which was in the moment, classified as the planet. In the time, Pluto was closer to the Sun as compared to Neptune due to its irregular orbit.

On 13 June 1983, Pioneer 10 became the first spacecraft that left the orbit of the main Solar System planets, effectively departing from what most people consider to be"the" Solar System, by crossing the orbit of Neptune The outermost of the solar system's eight planets.

On the 31st of March, 1997 NASA declared the end of their Pioneer 10 mission. Spacecraft was at an altitude of 67 AU from the Sun however, even at the cold and lonely distance, it could still transmit useful data beyond this point, the entirety of which was recorded the length of time that was possible.

Following the conclusion of the mission Deep Space Network continued to monitor the probe. Deep Space Network continued to monitor the probe for practice for the newly trained flight controllers. Finding high-frequency radio signal from deep space, particularly weak signals, isn't an exact science but it's a tiny part of an art. In the months that saw a decrease in signal strength, researchers employed the latest techniques of applied chaos theory to extract valuable information from a signal being increasingly dominated from background noise.

On the 17th of February in 1998 Voyager 1 surpassed Pioneer 10 in terms of distance from that of the Sun at 69.419 an AU. This makes Voyager 1 the farthest human-made artifact that is not located in our home. Voyager 1 was able to beat Pioneer 10 because the older spacecraft was moving over one AU slower that Voyager 1.

On the 2nd of March 2002, researchers could receive an uninterrupted telemetry transmission for 39 minutes via Pioneer 10 at a distance of 79.83 AU, but even the use of Pioneer 10 had to come to an end. In April of 2002, the scientists in Earth could receive 33 minutes of excellent telemetry however, that was not their final success in receiving telemetry. Following that day the spacecraft was no longer receiving usable telemetry via the probe.

On January 23 2003, the weakening signal was lost when Pioneer 10 reached a distance of 80.22 AU (12 billion kilometers) from Earth. Since that time it was discovered that it was discovered that the Deep Space Network was unable to discern any signals coming that came from Pioneer 10. Another attempt was attempted on March 4th 2006. However, the signal was not detected among the interstellar radio. NASA determined they had found that RTG units had stopped producing enough energy to run the transmitter. In the end that no attempts were made to get in touch with Pioneer 10.

Although there's no way to determine whether Pioneer 10 remains intact, NASA has predicted that, at the time of its launch on January 1 2019 the spacecraft was about 121.69 (AU) (~11.3 billion miles) far from Earth and is moving 12.04 kilometers/second (26,900 miles/hour) further away to the Sun. It means Pioneer

10 will gain an additional 2.54 an AU away from Earth as well as from the Sun each calendar year for the next five years. From this distance, the light of the sun will take 14.79 hours to get to the spacecraft. This means the distance of Pioneer 10 is 14.79 light-hours away from the sun. The sun will appear a little larger that the moon. Moon on Earth with a brightness of -16.6 that is about 35 times brighter than a Full Moon.

Based on the data available of the velocities and tracks, Voyager 2 supposedly surpassed Pioneer 10 in its distance from the Sun at the beginning of April 2019.

Pioneer 10's location is within the Taurus constellation, usually toward the giant red stars Alpha Tauri (Aldebaran). With the velocity of the probe the probe would need about two million years to get there. the current Aldebaran location. However, the fact is that all stars do not have a fixed position which means that in roughly 90000 years from now the probe is expected to be just 0.75 millimeters of K8V (deep dark, red dwarf) star HIP 117795 which is currently located situated at 87.3 light years from the Sun as well as Earth. It is the closest any of the interstellar spacecraft (both Pioneer and both Voyager probes) is expected to come to another star in the coming millions of years.

Today, a fully functional backup probe, coded Pioneer H, is on display in the National Air and Space Museum in Washington, D.C., within the "Milestones of flight" gallery.

On October 5, 1979, Pioneer 11 began its interstellar mission. The 25th of February in 1990 marked a day of red letters in the story of Pioneer 11. That day, Pioneer 11 turned into probe #4 that was able to travel over the borders of outer planets.

In 1995, the spacecraft did not have enough power to power its sensors which is why NASA took the decision to shut down the system in 1995. When expressing its appreciation for the successes in Pioneer 11, the Ames Research Center that was the one responsible for the mission, issued an announcement the following "After more than 22 years of exploration into the outermost regions in the Solar System, one of the longest-lasting and most productive missions to space in the history of science will be coming to an end."

On the 24th of November in 1995, researchers could receive brief telemetry signal that could be used for data, however that was the last time they spoke to the spacecraft. In 2002 when the spacecraft's signal was lost, it was too weak to be discernible in the noise of interstellar space.

As of the 30th of January of 2019, the probe was travelling at the distance that was 100AU (15 billion km) away from the Sun and was traveling at an incredible 11.241 kilometers/second (25,150 miles an hour) in the direction of the Sun. It is approximately

2.37 AU per year. Pioneer 11 is not headed to any specific star however, it is generally heading towards the north towards the Scutum constellation. Pioneer 11's closest interaction with another star over the coming few years is likely to be an orangish K-type dwarf, TYC 992-192-1. It will be roughly 928,000 years in the future. The closest approach is approximately 0.815 milliseconds.

Each of the Voyager probes have taken over Pioneer 11, and Voyager 1 is currently the furthest constructed by humans to its home planet.

Chapter 5: Venus Projects

As Pioneer 11 was on its journey to Jupiter towards Saturn, NASA launched the new spacecraft of the Pioneer program called The Pioneer Venus Orbiter. The following is information about its mission:

Also called Pioneer Venus 1, or Pioneer 12.

Date of Launch 20th May, 1978.

Launch vehicle: Atlas SLV-3D Centaur-D1AR.

Launch Location: Cape Canaveral LC-36A.

Cytherocentric Orbital Insertion Date: December 4 December, 1978.

Pericytherion: 181.6 kilometers.

Apocytherion: 66,630 kilometers.

Orbital Eccentricity: 0.842.

Orbital Period : 24 hours.

The Last Call: 8 October in 1992.

The probe contained these science tools:

Airglow ultraviolet spectrumrometer (OUVS) that detected scattered light and emits ultraviolet light.

An atmospheric drag experiment is used for studying the upper atmosphere and its volumetric density.

The charge-resistance analyzer (ORPA) which was used to study Ionospheric particle size.

Cloud Photo-Polimeter (OCPP) used to determine the cloud vertical distribution identical to Pioneer 10 and Pioneer 11 imaging photo-polarimeter (IPP) to fulfill their missions to Jupiter as well as Pioneer 11, which is the mission of Pioneer 11. Saturn.

The Electric Field Detector (OEFD) to study the solar winds and their interactions.

Temperature of electrons (OETP) used to investigate the properties of heat in the Ionosphere.

Gamma ray burst (OGBD) detector for recording gamma ray burst events.

Infrared Radiometer (OIR) to measure IR emission emanating from the Venusian air.

Ion mass spectrometer (OIMS) for analyzing the population of ionospheric Ions.

Magneticometer (OMAG) used to study the magnetic field of Venus.

"OnMS" stands for Neutral Mass Spectrometer (ONMS) used to measure the composition of the upper atmospheric.

Radio occultation test to characterize the atmosphere.

Radio Science atmospheric as well as solar wind turbulence experiments.

Solar Wind Plasma Analyzer (OPA) to measure the properties of solar wind on Venus.

The Surface Radar Mapper (ORAD) to determine topography and characteristics of surfaces. Observations were only possible in the event that the probe was located closer than 4,700 kilometers above the globe. An S-band of 20 watts (1.757 gigahertz) bounced off the surface with the probe analysing the echo. The resolution of the probe at periapsis measured approximately 23 by 7 km.

• Two radio science experiments with the aim of studying the gravity field of Venus.

In this way, a large amount of scientific equipment was utilized to research the vast Venusian atmosphere. When compared with Earth the pressure on Venus's surface is 92 times higher on Venus.

The surface mapping provided NASA an extremely detailed image of the way Venus is like under the thick cloud layer. On the highest level of clouds Venus can experience extremely high winds as a result of the huge difference between night and daytime temperatures. At the surface, however, Venus has virtually no winds at all, and rarely more than 10 km/h. This may be due to the fact that the atmosphere's thickness is extremely high in the upper layers, but it is more to have to do with the fact that temperature on the surface is almost the same everywhere at a blistering 462 degC. Like any student of meteorology, physics or sciences knows, it requires temperatures that differ to trigger winds to blow. Meteor craters that have been around for millions of years in Venus's surface Venus remain in perfect in their original condition due to the complete absence of wind, and the destruction that would be a part of it.

The next mission was Pioneer Venus Multiprobe, which included the following information about its mission:

Also referred to also as Pioneer Venus 2, or Pioneer 13.

Launch Date: 8 August in 1978. This was a bit two months later than the launch of Venus orbiter.

Launch vehicle: Atlas SLV-3D Centaur-D1AR.

Launch Location: Cape Canaveral Cape Canaveral LC-36A.

Venus Arrival Date: 9 December 1978, five days following the launch of the Pioneer Venus Orbiter (Pioneer 12).

The Last Call: Dec. 9 in 1978.

The spacecraft comprised four probes and a bus each of which was to be pushed into the earth's atmosphere via an heliocentric orbit. The bus exploded in the air at a height of around 110 km. However, the probes were equipped with heat shields and parachutes that helped them glide through the sluggish air. Instruments on board allow the probes to be able to test various aspects of the atmospheric environment as it travelled down.

The mission is represented in an artistic way. mission

The bus's image

"Large Probe," also known as the "Large Probe" was positioned on the dayside of our planet, just near the Equator. Another probe on the day side (called"Day Probe," or "Day Probe") was positioned just south of the Equator, and a little away from the Large Probe. Its "Night Probe" was positioned on the darker side of our planet and was about the same distance from the equator south than the "Day Probe." The "North Probe" was located about 30 degrees from 30deg from the North Pole of Venus, and was also located in the dark part of Venus.

The probes, with the exception of one ceased sending telemetry immediately after the point of contact on the ground. However, one probe, the Day Probe sending data for just over an hour following its collision on the ground. Surface temperatures varied between 448degC and 459degC. The surface pressure at four points varied from 86.2 bar to 94.5 bar.

While the probes soared into Venus's clouds the probes measured winds of around 200 meters per second (720 kilometers per hour) in the cloud's middle layer around fifty meters per second (180 kilometers/hour) in the cloud's base while only one centimeter per second at the top (3.6 km/hour).

Between the two missions researchers learned lots about Earth's twin.

Its Pioneer series of spacecrafts made up the most long-running program of NASA's relatively young existence. It ran between 1958 and the last encounter with a Pioneer probe in the year 2003 (Pioneer 10). This program gave about 45 years of support to the world.

The initial Pioneer mission failed and provided engineers crucial information that allowed them to develop more secure launch vehicles for subsequent missions, which included those of different programs. Although they were not perfect the Pioneer missions were all an important addition to the increasing amount of information about technologies for space.

The first Moon missions were followed by various programs that were more successful in discovering more about the Earth's closest neighbour in space. Pioneer expanded, pushing out beyond the Earth-Moon subsystem into interplanetary space to find details about the countless kinds of particles that make up what is normally thought to be the perfect vacuum.

Then, with Pioneer 10 and Pioneer 11 Pioneer 11, the program has earned its own name for being the first spacecraft that ventured out of the Solar System, offering close views of Jupiter as well as Saturn.

These two spacecraft are currently, two of the small number of spacecraft that will travel to interstellar space that is beyond the reach of our Sun's gravitational tyrannical grip.

Beyond the revolutionary results in the NASA's "Grand Tour" test run, Pioneer switched its final two missions to the closest neighbour, Venus - Earth's twin in terms of size and weight.

In the real world it was true that the Pioneer program was the basis to a highly successful start in the human race's Space Age.

Chapter 6: The Lure Of Mars

The first days of July 20, 1976 was filled with nervous tension as engineers and scientists eagerly awaited information to be returned to Viking 1. Viking 1 spacecraft. Humanity's robotic ambassador was the distance of 216 million miles, this meant signals could require over 19 minutes for them to get on Earth. If something went not right, the scientists would find the cause until long after the time was up to take action.

It was the Jet Propulsion Lab (JPL) located in Pasadena, California had come in the hopes of scientists and engineers who worked for over a decade working on the JPL project. In the early hours of early morning had been peaceful in this part of Los Angeles County, but when darkness gave way illumination, the darkness was replaced by a throng of scientists, journalists as well as administrators, technicians and others. A waning moon shone across the San Fernando Valley and Santa Monica

Mountains to both sides of the illuminated campus. The other side of the campus was the shadow of the San Gabriel Mountains. Mars was set some several hours prior to the sun's setting.

Viking command and control room was tiny room that had glass walls permitting the rest of the group to peer at the men who were hunched in a circle around the console as they watched the stream of constantly changing data displayed on TV screens. Cameras across all over the Jet Propulsion Lab campus projected the image of the room. Behind those men and their consoles was the glass-walled offices that belonged to Viking Project Manager Jim Martin Jr. He had been in control of the project throughout the several months. With up to $1 billion on the line for the two Viking spacecrafts, Martin Jr. wanted nothing to go wrong that would ruin the result.

Martin

Following the modest rockets of the lander were able to significantly slow its speed due to its deorbit burn everybody waiting for the lander at JPL had nothing else to do other than stare at the screen when the spacecraft was speeding towards the ground in a lengthy thin circle. Al Hibbs, senior advanced mission planner, took on the role of "voice for Viking" when the spacecraft accelerated towards. In the long hours of wait, Hibbs attempted to keep everybody updated with straightforward positive and encouraging information. "So far all that was suppose to be happening has happened...has been completed and ahead of schedule. We're rapidly getting closer to the Martian surface. Mars ..." Then He would also not forget the distance remaining: "Only 7,048 miles to travel."

at 4:53:14 a.m. (PDT), Hibbs informed everyone that "Viking is likely to be visible at some point or other." The public wouldn't be able to know in the next 19 minutes.

A different voice coming from Mission Control added altitude and speed data as if it came from across the huge gulf of space between the two worlds "Speed 4.718 meters per second. Altitude 98,707 meters."

at a height of around 30,000 metres at a height of 30,000 meters, at a height of 30,000 meters, Martian atmosphere had gotten dense enough to provide substantial braking force. This slowed the ship's descent down to 3000 m per second. At 27,000 meters, the descent was slowing even further by 1,820 meters per second.

At a distance of 22,800 feet, the ship was descending to 982 meters in a second.

5:09:50 in the morning, JPL received the delayed announcement that the landing craft has deployed its parachute. It then decreased speed to 709 m/s.

At 5:11:27 AM, 1,463 meters, 54 meters per second.

Once the spacecraft was at the altitude of 1,400m and its final engines for descent came on to slow the speed of its return to the Earth's surface.

Then, at 5:12:07 a.m. PDT, Mission Control received the signal that the Viking 1 lander had indeed arrived on the planet of Mars and arrived 45 minutes earlier than the local dawn and the majority of this Southern California megalopolis was still asleep. Martin shared his congratulations with the other members of the team, as well as in JPL's auditorium JPL auditorium, reporters as well as public relations officers from NASA were able to share a glass with Cold Duck to celebrate. A chorus of cheers filled the morning air on all over the JPL campus.

While when the Viking 1 spacecraft was landing on Mars The president Gerald Ford called Martin at JPL. Martin responded that Martin was "busy at the moment," and that the president could call him back within three hours. Martin's requests and Martin went back to the task in hand. The manmade object reached an undiscovered globe, one with an atmosphere and maybe life.

The interest of people around the globe in Mars is traced to a few thousand years' of studying the planet. Since the days of the astronomers from ancient times, Mars is a significant element of popular art and science. Astronomers from the past followed Mars its movements for over 3500 years ago. It was in part because ancient cultures believed that the planets and stars within the sky were of significant importance in their religion.

The curiosity for Mars and the potential for intelligent life there reached a fevered level in the 17th century, as scientists made precise observations using telescopes. When astronomers noticed different color spots across Mars's surface Mars and the appearance of an ice cap that was polar The similarities with Earth (and consequently the likelihood of living things) appeared to be higher. The 19th century was when Astronomers were able to see the appearance of clouds and storms on Mars.

When scientific studies of Mars got more specific, Mars became a pop popular. The late 19th century was the time when Mars became a popular subject the concept that there may be small green Martians in Mars Red Planet still had legs within the scientific community and even science fiction writers such as H.G. Wells published numerous accounts of encounters to Martian aliens. Wells' The War of the Worlds recounts the tale about the assault on Earth by Martians And while human beings do not stand a chance against Martians Martians however, they are Martians can't compete with bacteria that eventually destroy those aliens that invaded the Earth. Wells' Martians had apparently accomplished interplanetary travel, but they did not have using antibiotics.

Wells wrote a book that wasn't real, but it felt quite real for a few viewers on the radio on the 30th of October the 30th of October, 1938. On that day, Orson Welles read The War of the Worlds in a fake news program featuring news bulletins updated regularly that informed listeners about how far they were getting through The Martian invaders. The listeners who had not read the book were misled into thinking that a war was occurring. The incident is well-known in the present for creating a sense of panic for those listening to radio. However, the magnitude of the panic is not clear. The most definite result is that it made Orson Welles famous. That was before he acted and directed on screen in Citizen Kane.

Viking's technical Challenges

The initial step of the process of planning a Mars mission is achieving the orbit of Earth, which involves the use of powerful rockets for launching a spacecraft. They propel it to a point where the spacecraft travels at a speed of nearly 18,000 miles an hour, which is what it takes to reach the orbit of Earth. Speed, accuracy and the technology required to reach the orbit safely is difficult to attain even now, and certainly not when it was 1960. The two most devastating disasters for NASA that resulted in the loss from Space Shuttle Challenger in 1986 and the Space Shuttle Challenger in 1986 as well as Columbia in 2003, Space Shuttle Columbia in 2003 were caused by problems caused by launches.

After the spacecraft has been placed in orbit, it'll eventually come back into Earth in the event that it fails to quit the orbit. The rate required for the object to depart orbit is called the escape velocity. In order for an object to escape velocity that allows it to leave Earth's orbit the spacecraft has to be travelling at a speed of 250 miles per hour in addition to the fact that the speed required for reaching escape velocity is much greater than the speed required to get into the orbit of Earth, spacecrafts have to utilize rockets that have multiple "stages" which push forward and accelerate at different moments throughout their journey. In the case of, for instance, the Saturn V rockets used for the Apollo missions could utilize three stages in order to propel an orbiting spacecraft to Earth as well as the third stage could be activated to propel the spacecraft until it reaches the speed of escape.

When engineers were creating rockets capable of propelling objects to escape velocity in the 60s, mathematicalists as well as scientists had to work out what position Mars was in its orbit, and the time it would be there. As Mars and Earth have orbits that differ from their respective orbits around the Sun, Mars is in the same location in relation to Earth as well as the Sun just once in 780 days. This is referred to as synodic periods that is essential to calculate it precisely in order so that the spacecraft can be taken by Mars its orbit. Spacecrafts will orbit the Sun all the way therefore it's not as easy as shooting an spacecraft along a straight direction towards where Mars is. Making it all work is extremely complicated mathematical calculations. Therefore that when NASA decides to launch the Mars space mission, the agency gets one shot to make the job done right or there could be delays of almost all synodic time.

In case figuring the entire process wasn't difficult enough The spacecraft must also be able to get into Mars the orbit following its escape from the Earth's orbit. Because Mars's gravitational pull is only half that of Earth's, the spacecraft has to slow in the process of approaching Mars which allows Mars Red Planet to "capture" the spacecraft, and then draw it into. If the spacecraft was to remain in the same speed for escape from Earth's orbit it will in essence "bounce" away from Mars the gravitational field. It would then speed right through the red planet.

The technologies that contributed to the success of the Viking Program can be traced in World War II, when Nazi Germany employed rockets for bombing England. When the last rounds in World War II were fired and the reconstruction of Germany and Europe started in the aftermath, Europe and Germany, Western Allies and the Soviet Union both tried to get access to the top scientists of the Third Reich and engineers, particularly those working on the fields of rocketry, missile technology as well as aerospace research. Naturally this was not an easy matter due to the fact that most of the German scientists weren't just actively Nazis however, they had assisted in to help the Nazi army to conquer all over the globe. However in the post-war period in the late war, it was clear that the Western Allies formed a clear image of the Soviet state. While they were compelled to unite with the Soviets however, the West began to realize that Communist Russia as a aggressive totalitarian power and therefore a significant danger to a free Europe.

Both Western Allies and the Soviets had knowledge of Hitler's V-2 rocket project that was the precursor of ballistic missiles, as well as the race to space. They both recognized the enormous significance of these technologies and sought to safeguard the benefits of these technologies in their own interests. When the Soviets were contemplating expansion after the "Great American War" as well as the U.S. military came to recognize that their supposed allies from now would become tomorrow's enemies those who had knowledge of V-2 rockets as well as others Third Reich military technology programs began to be seen as essential elements in the upcoming NATO to Warsaw Pact standoff.

This led to the US-led "Operation Paperclip" on the Western part, which resulted in German scientists placing their skills to the use by their fellow NATO members, including the U.S. and other NATO members. Operation Paperclip aimed not only to gain the benefits of German technological advances to America but also to protect them from the United States but also to keep them from the hostile Soviets in the manner that General Leslie Groves enunciated: "Heisenberg was among the top physicists in the world as of the period of the German split, he was valued more than the ten divisions made up of Germans. If he had fallen into Russian hand, he might be a valuable asset to the Russians."

The Western strategy, though selfish generally met with a consent on German scientist's part. Contrarily to the Soviet solution to Paperclip, Operation Osoaviakhim, employed the implicit threat of torture, imprisonment and even death, one of the most common tools employed by Stalinist Russia for coercing help from German engineers and scientists following the outbreak of war. They paid rich dividends for the Soviet government in the sense in achieving, at a minimum, technological parity with its west-bound competitors.

to say that Operation Paperclip had a profound influence on the Cold War and American history could be an understatement. One of the most famous examples of the mission's "success" is the Wernher von Braun. Von Braun was a part of an SS branch. SS that was involved in the Holocaust He would be known for being"the "father for rocket technology" and would enthral all of the world with images of winged rockets as well as space stations in an "new" Manhattan Project, one which NASA was to later take over. In addition to the ballistic missile weaponization which developed through the Cold War, von Braun's knowledge was utilized in one of the oldest space missions in history. NASA was also required to create rockets that could first launch an orbiting spacecraft within the

Earth's orbit before launch it towards the Moon. The Soviets were unable to develop rockets capable of this task. However, due to von Braun, NASA got it just right with its Saturn V rocket, which is still the most powerful launcher NASA has ever employed.

Von Braun

Following World War II, America formed The National Advisory Committee for Aeronautics (NACA) and, however, after the Russians launch Sputnik in October of 1957, the federal government decided to take the risk seriously enough to approve the establishment of National Aeronautics and Space Administration (NASA) in the year 1958. In the following three years the president Kennedy was to promise to place an astronaut on the Moon at the close of this decade.

A human-powered mission to the Moon will require the most cutting-edge rockets for spacecrafts to the Moon as well as ensuring an efficient reentry of the spacecraft back into the Earth's atmospheric space. In 1961, when Kennedy presented his vision in 1961 NASA didn't yet have the technological or scientific knowledge necessary to make the return of spacecraft a reality to Earth, as well as the spacecraft or rockets were not in existence at the time.

NASA was unsure of what their plans were for a spacecraft designed to travel up to the Moon and return. From the start NASA's engineers thought it would be the most efficient to launch the spacecraft directly into the Moon and then equip it with boosters for launching it back down to Earth. The method, known as "direct ascent" appeared feasible to NASA due to the fact that the Moon did not have an atmosphere and therefore, ascent from the Moon quite simple.

There was a small percentage of NASA employees favored different designs. A design known as "Earth Orbit Rendezvous" could have tried to join several pieces to form a spacecraft on space. It is believed that the "Earth Orbit Rendezvous" was used to construct parts of the International Space Station, but it was not within the capabilities of NASA in the early 1960s.

A different group of supporters favored the idea that was later referred to in"Lunar Orbit Rendezvous. "Lunar Orbit Rendezvous" mission. Lunar Orbit Rendezvous required having an rocket launcher and with three main parts: a command module, command module, and a lunar module. Three of the modules will remain in place until the spacecraft started moving around the Moon after which the lunar module was released and dropped to the Moon's ground while the Command/Service module circled around the Moon. People who advocated for this kind of mission claimed that this was the ideal choice since it required landing on the smallest amount of weight at the Moon. In the year 1962 NASA had concluded that the Lunar Orbit Rendezvous mission was the ideal choice.

After NASA was able to decide on the Lunar Orbit Rendezvous for the Apollo missions, they needed to develop and build the spacecraft and rockets that could be built in modular fashion. At the outset, NASA faced a major logistics issue; any lunar landing requires several crew members, however NASA did not have an spacecraft with more than one person. For this purpose, NASA designed a Command/service module. It would house vital elements like water, oxygen as well as power. three astronauts would remain inside the command module for the liftoff, orbit, and reentry. The lunar module that is ultra-light, that was developed for ascending and descending towards and away from the Moon can accommodate two astronauts. When it was time to ascend from the Moon the lunar module could dock together with the command/service module.

The design of the spacecraft was not the toughest aspect. NASA was also required to create rockets that could first launch the spacecraft's modular components into the orbit of Earth, before sending it towards the Moon. The Soviets battled for years in developing rockets that could handle the challenge, however NASA succeeded using their Saturn V rocket, which is still the most powerful launcher NASA has ever employed.

Chapter 7: The History Of Nasa

The National Aeronautics and Space Administration (NASA) was founded on the 1st of October, 1958 but the story had begun taking shape for years prior. In the U.S. Department of Defense decided to explore rocketry research, space exploration and atmospheric sciences at the end of the 1940s as a way of making certain that the United States would be a technological pioneer globally. In approving the orbit of an observatory satellite for scientific research to be part of the United States' contribution to the International Geophysical Year, President Dwight D. Eisenhower advanced this goal (July 1st, 1957 until December. 31st in 1958). The 67-nation alliance was created to study eleven Earth sciences, and also collect data on Earth. However, when it was announced that the Soviet Union announced its own plans to put a satellite into orbit and contribute to the space race, American interest rapidly waned. According to their plans to launch a satellite, it was the Soviet Union launched Sputnik 1 which was the first artificial satellite on the 4th of October in 1957, in advance of United States on its own initiative. This began the what was later to be called"the "Space Space Race" between two powerful nations.

On the 31st day of January in 1998, Explorer 1 was successfully launched by the United

States, but the Soviet victory already had triggered protests from patriotic Americans and their supporters, who demanded an increase in funding for aerospace programmes as well as technology. The Committee on Space and Astronautics was created in Congress US Congress on the 6th of February in 1958 to develop laws for a national space program. The President Dwight D. Eisenhower signed the National Aeronautics and Space Act into law on the 29th of July in 1958. It was signed a couple of months after. Then, two months later NASA began operations.

The good news is that the company was not required to create by scratch since it was already incorporated into it into the National Advisory Committee for Aeronautics that was in existence since 1915. It was an organization that brought more than 8,000 people, a $100 million budget, three research laboratories (Langley Aeronautical Laboratory Ames Aeronautical

Laboratory, and Lewis Flight Propulsion Laboratory) and two test facilities that were smaller. The other organizations were also quickly connected, like the Space Science Department of the Naval Research Laboratory department located in Maryland and The Army's Jet Propulsion Laboratory managed by the California Institute of Technology, as well as the Army Ballistic Missile Agency in Huntsville, Alabama, where Wernher von Braun's group of engineers had already been engaged in large-scale rocket development.

Following the year, NASA embarked on human space exploration initiatives, such as Project Mercury, a single programme for astronauts to assess whether humans could live in space. Project Gemini, a dual project for astronauts that aimed to perform activities in space, such as rendezvous, docking and docking of vehicles and spacecraft as well as Project Apollo to land on the moon and study it. America was the only country to walk to the lunar surface in July 1969 accomplishing a feat which was previously believed to be science fiction.

The development of the International Space Station (ISS) along with the launch and operation of the Hubble Space Telescope, as and numerous advances in the fields of biotechnology and biology, earth and space research physical science and technological development, which included a variety of consumer products that were created thanks to NASA technology. These were enabled through NASA's Space Shuttle Program, which was in operation from 1981 until 2011.

In the present, NASA continues to support the ISS by implementing the Commercial Crew Program while also working on deep space exploration through the creation of Space Launch System, or SLS and will one time transport astronauts to distant spaces such as Mars asteroids, asteroids and even beyond. NASA is currently operating seven field centres, ten field center laboratories for testing and research, and has its main office located in Washington, D.C. NASA employs more than 18,000 employees as well as a large number of contractors from the government.

What is the ARTEMIS PROGRAMME Of NASA?

NASA is in the process of coordinating a variety of space missions in the Artemis program.

Three Artemis missions currently in operation. Artemis 1, an unmanned testing flight that flies around and over the Moon is scheduled to launch in September 2022. (Postponed until 28th August). Humans are expected to travel the most through space in the history of mankind in space on Artemis 2, a crewed mission to the Moon. and Artemis 3, a mission which will bring the first female, and also the first human from a different race to the Moon for research on lunar surfaces for a period of one week.

Artemis 3 is the US space agency's first moon landing by crew operation after Apollo 17 in 1972.

The goals for the future of NASA are more challenging however, the Artemis space mission is primarily focussed on exploration of the moon.

NASA is planning to send a team to Mars at some point by using the research and technology generated through the Artemis space missions.

A brand new space station is expected to be built in lunar orbit as part of NASA's lofty "Moon to Mars" program as well as a base where the moon could eventually be habitable.

The reason why the program is called"ARTEMIS"?

Artemis is a legend of the Greek goddess associated with the Moon as well as the sister to Apollo. This connection to Apollo's missions that sent the first human beings to the Moon fifty years ago is evident. In the meantime, Orion is the name of the spacecraft that is currently in construction. Actually, Orion is among the most famous constellations in the night sky. In the mythology of classical times, he's Artemis the hunter.

Artemis was a significant god in the ancient world of Greece that was revered in the early millennium B.C., if not earlier. Her mother was Zeus The god of the sky, Zeus. He was the most powerful Olympian god who was the ruler of all the earth from the top of Mount Olympus.

Artemis who is regarded as the goddess of the animals and wild, also has inspired conservation initiatives for the environment, where the goddess is considered to be an example of a woman able to exert power through caring for the earth.

Although it was true that the Greek Artemis was powerful and brave, she was not always a good and loving person especially towards females. Her impulsiveness was often utilized to justify a woman's untimely death, particularly when having a baby. The mythology of this goddess is now fading with the passing of the years. Artemis has been regarded as an emblem of femininity and self-reliance in the wake of feminism's rise.

NASA's mission has been named for mythological creatures. A number of launch systems and rockets were named in honor of Greek gods of the sky including Atlas and Saturn who's Greek title is Cronos since the 1950s.

Atlas As well as Saturn were titans, and not only gods. Titans in Greek mythology. Titans from Greek mythology are the unruly natural forces that are primordial to nature They also represent the immenseness of the space-based exploration. The Titans were renowned by their immense force and strength, however they also were rebellious and dangerous. They were ultimately defeated by the Olympians which represented the civilisation in Greek mythology.

After the advent of human space flights, NASA began naming missions in honor of Zeus's children and a connection with the skies. Mercury was the name of the program. Mercury program, running from 1958 until 1963 was named in honor of Hermes who was the Roman god of messengers who used his wings to travel through Olympus, Earth, and the world beneath.

The three-year Gemini program that began in 1963, was the Gemini capsule, which was designed for two astronauts. The program was named for Zeus twin sons Castor and

Pollux also known as Dioscuri in Greek and were portrayed into the sky as the constellation Gemini. When it came to Greek or Roman art it was common to depict them wearing a sterno above their heads.

Names like Columbia, Challenger, Discovery, Atlantis, and Endeavour were chosen in order to convey an energy of creativity throughout the space shuttle program that was in operation from 1981 until 2011.

NASA has paid tribute in honor of the Apollo program that was in operation from 1963 until 1972. It was the first program to land humans in space on the Moon in 1969. Artemis will take over where her twin brother had left in the 50 years that followed and will usher in a varied era in humans in space.

Chapter 8: A Few Facts On Artemis Fun Facts

ORIONS are going further than any OTHER human-powered vehicle on the planet.

When out in space Once in space, the SLS Interim Cryogenic Propulsion Stage will give the spacecraft the necessary moon push before it is able to fly on its own. With the help of trans-lunar injection, Orion will travel 280,000 miles (450,000 kilometers) far from Earth. If the system's checks are successful, Orion will be tasked to settle into an orbit that takes the spacecraft just sixty-two miles (100 kilometers) over the lunar surface when it is at its lowest. After that, Orion will use the moon's gravity to reach the depths of space than before-to-date spacecraft rated for human use, and will travel around 40,400 miles (64,000 kilometers) away from the moon. Orion is further away than the three astronauts from Apollo 13 that had a slightly altered orbit around the moon back in 1970. They flew into a deeper space than other Apollo missions. (The astronauts faced an issue with their spacecraft and had restart their engines many times before they could get back in the right direction for a secure returning home to Earth.)

The SPACECRAFT will be flying for over a month.

Orion is expected to spend between 39 to 42 hours in the space according to the time of launch. As an example, launches in September 29th or August 5 could have a 42-day space mission while a launch on September 2 will begin the 39-day mission. (The variation result from orbital dynamics as the Earth and moon are constantly in motion relative to one another.) NASA extends its orbital period to prove a point. NASA wants to push the spacecraft beyond its limits in order to make sure it's ready for human use in the future, going as that it's "go" in order to cause failings that wouldn't be acceptable to a person who is in the spacecraft.

According NASA's Artemis 1 mission manager, Mike Sarafin, "We're trying to minimize the risk of flights with crews, and so we're prepared to take on more risk with an uncrewed test flight." He said using this method would give NASA with an "lean forward" time to ensure that they are ready to host humans on the next flight.

THE ROCKET HAS NEVER BEEN LAUNCHED BEFORE

SLS is the strongest rocket that has ever been built, and it is still firmly under the gravity pull of Earth. Two boosters will be supported by the main stage that is more than 200 feet (61 meters) high. The engines of the core stage are powered by 730,000 Gallons (2.76 million of liters) from supercooled liquid oxygen as well as hydrogen. (By by the way, these engines are from NASA's space shuttle program which was terminated in the year 2011.) The rocket has been through several tests, such as a long-term "wet formal rehearsal" during June 2022 where the rocket was fuelled and carried most of the way through a virtual countdown.

The rocket has passed the tests required to be accredited for flight according to NASA despite the hydrogen leak as well as additional technical problems. There are still issues to be resolved, the experts said, can be addressed at the "hangar" (the Vehicle Assembly Building) prior to the launch.

ALEXA will be on BOARD

One of the passengers onboard Orion spacecraft Orion spacecraft is the Amazon's Alexa. The voice assistant is well-known across Earth due to its capability to control a variety of devices for personal use, such as phones, speakers, as well as Internet of Things gadgets. Callisto will be demonstrating technology is expected to aid in the mission to the moon on Artemis 1. With Orion manufacturer Lockheed Martin, Callisto will demonstrate the latest technology in voice, artificial intelligence and tablet-based video demo. The program also features Cisco's Webex video conference software.

The technology was first considered in the month of January, 2022 Lisa Callahan, vice president and general manager for commercial civil space for Lockheed Martin, said in an announcement that "Callisto will showcase a new technology that will be employed in the future, allowing astronauts to become more independent in their exploration of deep space."

A HUMAN SPACECRAFT running on the power of solar will go to the moon.

When using spacecrafts like the International Space Station, humans utilized solar energy previously in space, before the safe confines that are Earth orbit. (The

Apollo moon astronauts, that operated during the 1960s and 1970s depended on fuel cells to power their systems.) The dynamic of this will change due to Orion with the help of solar power near the moon. This is because the European Space Agency (ESA) provides the oxygen as well as water and energy to Orion. Orion spacecraft. This will all be powered by electricity derived through solar energy taken up through 3 solar arrays. What type of energy? It's quite a lot. According to the power source, to ESA it is comprised of three panels in each one of four wings that can supply power to two standard European houses, while providing 120 volts to computer, experiment equipment and various other devices.

SNOOPY is on the board.

The characters of characters from the "Peanuts" comic book will appear on NASA's next spacecraft Artemis 1, continuing a tradition that has been a longstanding NASA tradition. The cute Snoopy the Beagle will serve as a soft toy, as well as a "zero gravity gauge" for Orion. Snoopy is best known for the moniker of the lunar module Apollo 10 used to practice landings on moons during lunar missions.

According to the company, Snoopy will be outfitted with Orion with the exact orange NASA jumpsuit, as will the future Artemis astronauts. They will also be wearing the Silver Snoopy pins, a popular award that has been awarded to top performing NASA staff, will accompany us on our journey. The award also pays tribute to "Peanuts" creator Charles M. Schulz (who passed away in 2000, aged of 77. His last strip appeared the following day.)

A long-winded statement by NASA, "a pen nib of Charles M. Schulz's Peanuts Studio will travel to Artemis I, wrapped in the space-themed comic strip part of a set of memorabilia to travel onto the Orion spacecraft. NASA is studying each interaction it has been involved in with "Peanuts" throughout the past.

Millions of names of SPACE FANS' NAMES will be found in deep space.

NASA offered space-lovers the chance to join the Orion spacecraft. This was a continuation of a traditional space-based tradition that has been in place for a long time. Anyone could make entries on the agency's website. The names were then loaded on an external flash drive, which was then stored on the ship.

A team of 30,000 has been credited with helping to improve the quality of all submissions of which were nearly 3.4 million. These people helped Orion to accomplish its mission. In order to "show appreciation to dedicated personnel who gave their time and energy in the space mission." NASA said those names were imprinted on microchips.

As per the government agency the most significant contributor categories were people from NASA, ESA, industry as well as suppliers. The categories ranged from mission-related operations to a variety of flights programs.

EXTRA SCIENCES WILL BE RECOVERED by 10 cubesats.

A cubesat group consisting of 10 are scheduled to be launched in various missions that will study the moon, the sun as well as near-Earth asteroids and will be transported through Artemis 1. (Cubesats are tiny spacecraft which have been utilized for various purposes throughout Earth through time, such as observations, communications as well as other applications.) A majority of these tiny spacecraft is designed to be used to study the lunar surface, however there are a few that serve other functions. One of them in its class is NEA Scout, which will explore an asteroid for 2 years. This mission, powered through a solar sail will be able to closely examine an asteroid. (The specific target of the mission is to be set after the date of launch.) A camera called NEACam can take images with stunning 20 megapixel resolution. These will be sent the images back to Earth in the event that the spacecraft is 65 to 75 miles (100 to 120 kilometers) away from the spacecraft.

3 MANNEQUINS WILL FEEL RADIATION

Orion isn't going to have anyone aboard, however it will be equipped with three mannequins which look human-like and were supplied by NASA as well as Europe.

Two torsos of mannequins (also called Phantoms) were loaded on this spacecraft, by Matroshka AstroRad Radiation Experiment (MARE) scientific team. It also is comprised of scientists of DLR (the German space agency). Both spacecrafts, Helga and Zohar, have 5,600 sensors in them to determine the amount of radiation. Zohar is also wearing an AstroRad radiation-resistant vest in order to determine the impact the vest has.

The "moonikin" with its name Apollo 13 engineer Arturo Campos acts as the third "crew members. The moonkin features vibration, acceleration, as well as radiation sensors, which determine the effects of space stress on the person in launch as well as space flight and landing.

THE MISSION WILL END WITH A SPECTACULAR SPLASHDOWN

The Orion mission that has been erratic is set to come to an end by an unpredictable return. When it is close to Earth it must take off the service module in order that it is able to operate with batteries throughout the splashdown and re-entry phase. In contrast to normal missions with the low Earth orbits, the return back to Earth begins at an altitude of around 400 feet (122,000 meters) and will be extremely fast.

In a rate of 24500 miles per hour (39,500 km/h) The spacecraft is expected to hit the earth's atmosphere, creating deceleration forces that can be as much as nine times the force of gravity (a normal ISS launch could produce three decelerations, in comparison). The outer surface of the spacecraft could reach temperatures that range from to 5,000°F (2,800 temperatures Celsius).

In over the Pacific Ocean, Orion will penetrate the sound barrier and signal to the teams that splashdowns are just close by. It will release an insulating thermal shield and then slow down using parachutes and drogues. It is expected to have dropped by a factor of 32 times that of sound until the moment it lands (Mach 32).

Chapter 9: Nasa Project Artemis

NASA hopes to bring its Artemis 1 moon mission off on the right track despite the recent issue. The vital mission, Artemis 1, whose first launch attempt on the 29th of August 29failed due to technical issues and is scheduled to be launched on the weekend of September 3. This announcement was made by the agency on August 30.

If everything goes to schedule, Artemis 1 will launch at 2:07 p.m. EDT on Pad 39B at NASA's Kennedy Space Center (KSC) located in Florida in a time span of two hours (1817 GMT).

The NASA Artemis program, as it's name suggests, is designed to ensure a permanent human presence in and around the moon before 2020. Artemis 1 will be the first mission of the program. In addition, it marks the launch event for NASA's huge new Space Launch System (SLS) rocket that will transport the unmanned Orion spacecraft on an extended voyage to lunar orbit, and then back.

The shake-out cruise was planned to depart. But, as the countdown began participants from Artemis 1's Artemis 1 team observed that one of the four engines which drive the SLS main stage was not getting to the proper low temperature of around minus 425 degrees Fahrenheit (minus 250 degrees Celsius) -just prior to the activation. As per mission team members, this process of thermal conditioning stops shocks when engines are lit up through "bleeding" into supercold propellant liquid hydrogen. As per John Honeycutt, manager of the SLS program at NASA's Marshall Space Flight Center in Alabama the engines 1 2 and 4 were quite close in the countdown. However, engine 3 was away from the mark with a temperature of minus 38 temperatures (minus 300 Celsius).

The Artemis 1 team was not able to identify the problem in a timely manner and this led to the cancellation of the attempt to launch. However, Honeycutt along with others on the team think they are in control as of now. They think Engine 3's defective temperature sensor is the culprit. Sensor readings indicated it was possible that Engine 3 was receiving the appropriate quantities of hydrogen liquid during the bleed, Honeycutt said. Removing the sensor could likely require rolling over the Artemis 1 stack off Pad 39B before returning to the massive KSC Vehicle Assembly Building, according to Honeycutt as well as other people in the meeting. It is believed that the Artemis 1 team does not think it is needed in the moment but instead is planning to launch the sensor again.

The team plans to introduce a few modifications in the countdown schedule that include initiating the process of cooling down by 30 to 15 minutes before the time of it was the last time. In addition, as Honeycutt said and his team, they'll continue to analyze the data and laying different scenarios in the next few days in order to determine if the approach currently in place is rational and sensible. It is imperative to keep analyzing the data, he stated. "Assuming we're right that using the Engine 3 bleed-temp sensor won't give more results, we'll need to develop some flight logic.

In an "wet dress rehearsal" an exercise that simulates launches and testing of fueling that helps the testing of a rocket before its initial flight. These problems are often resolved. In the spring of this year was no exception. The Artemis 1 team made a few wet dress efforts at Pad 39B However, they encountered a variety of technical problems and missed a few actions. Wet-dressed attempts were unsuccessful in putting Artemis 1 in the "engine bleeding" setup that led to the problem.

Based on Mark Berger, a launch weather analyst for the U.S. Space Force's 45th

Weather Squadron the latest forecasts indicate a possibility of rain and storms across the Florida's Space Coast. Berger declared that there's 60% likelihood of being a weather-related incident in the Saturday launch window in the press conference that evening. It's likely that the weather will improve within the timeframe. However, he did affirm his faith, and gave Artemis 1 a chance to begin its journey.

Why is NASA RETURNING TO THE MOON?

Through Artemis, NASA wants to explore the Moon and "stay there" instead of merely trying to recreate the successes that were made by NASA's Apollo missions. Although the primary goal remains to bring astronauts returning to the Moon in the middle of the next decade, this requires looking at possibilities of setting up bases on lunar orbit, as well as on moon's surface.

The primary objectives of NASA missions comprise

Equal opportunity: NASA's highest objective is to bring the first woman, and also the first human being of color to the moon.

Technology Spacesuits and rockets the latest technologies being developed are designed to allow for the development of deep-space missions.

Partnerships First major joint ventures with for profit companies such as SpaceX as well as Boeing are part of the Artemis program.

Permanent presence for a long time: Compared to Apollo 17 crew's three-day stay on the moon's surface Artemis is planning to create an infrastructure to allow visits that run for weeks and even months.

10: Nasa's Ten Very Best Achievements

The FIRST U.S. Satellite was an Explorer 1.

The moment news about Sputnik's achievements got to the Jet Propulsion Laboratory (JPL) and eventually became known as the NASA Jet Propulsion Laboratory, they began to develop the satellite that would go on to launch following it. Explorer 1 was completed by the JPL within just three months.

The satellite went out into space by a rocket and came with instruments for the analysis of cosmic rays that strike the orbit of Earth. Explorer 1 weighed 30 pounds and measured 6.25 inches (15.9 centimeters) in size and was eight feet (203 centimeters) in length (14 kg). Explorer 1 recorded cosmic radiation that was reflected off its surface when it circled the planet 12 daily with an altitude that ranged between 1.563 miles (2,515 kilometers) up to two hundred miles (354 kilometers) higher than Earth.

The tiny object changed our knowledge of earth's atmospheric conditions. Following the time that Explorer 1 arrived in space the spacecraft began to collect information regarding cosmic radiation. The data received from Explorer 1 revealed cosmic ray activity that was less than what scientists had anticipated. Scientist James Van Allen proposed that the reason for this was interference to the cosmic ray detector onboard. He believed Explorer 1 had passed through an unidentified radiation belt and inundated the onboard instruments by charged particle.

Van Allen's theories were confirmed with the help of data from a different satellite, which was put into orbit just two months later. At that point, the radiation belts that surround the Earth were recognized by scientists. After 58,000 rounds of orbiting the earth, Explorer 1 dropped into the atmosphere, and then burned on March 20, 1970.

Twenty years after the launch of a satellite revealed findings that go far over what could be found in the textbooks. Space photos with HD resolution could be transmitted to us through the payload on the spacecraft.

THE UNIVERSE UNVEILED by The Space TELESCOPE HUBBLE

Prior to the year 1990 in the year 1990, light telescopes that were at ground level offered the vast majority of our perspectives of space. Even though the pictures weren't exactly crisp, they were interesting, but the telescopes could not see enough distance to provide us with the images the astronomers had hoped for. In spite of all the clouds or water as well as gas vapor, the atmosphere of Earth can distort light from other planets which makes it hard to get clear pictures.

Solution: Place telescopes in the opposite part of the air, and light can freely travel to distant objects, and then return. It is believed that the Hubble Space Telescope (HST) could help in this. Though it's not the only space telescope but the HST is now considered to be one of the greatest technological breakthroughs in the history of science. This instrument, with its name after Edwin Hubble, has examined over 40,000 diverse celestial objects, and also performed more than 1.5 million observations of astronomy.

It is still able to provide the people of today crystal clear, fascinating views of the universe. Furthermore it is the reason that the HST continuously promotes new discovery. Nearly 15,000 papers in the field were based on information taken from the telescope.

In the mid-late 1970s NASA along with the European Space Agency worked together to construct the Hubble. The Hubble was originally scheduled to go into orbit by 1983. However due to delays with the construction of the Hubble as well as the political repercussions of the Challenger incident in 1986, the launch was delayed until around 1990.

Now we can look at the expanding universe in ways that we would not have imagined before because of Hubble Space Telescope. Hubble Space Telescope. Its resolution ranges from 10 to 20 times higher than what you can get from a standard ground-based telescope. A technological innovation made its images accessible to scientists as well as people who are not scientists. With the advent of the Internet the public has enjoyed the freedom of sitting at home and gaze upon the sky in all the high definition, vibrant color. For anyone curious, Hubble made the world billions of light-years away from Earth accessible.

Presently, Hubble collaborates with other excellent NASA telescopes to improve our knowledge of the universe. Prior to 2000, the Administration announced a brand new space telescope that captures breathtaking images of the universe making use of X-rays instead of visible light.

On CHANDRA The X-Ray, a HIGH-ENERGY UNDERSTANDING OBSERVATORY

Most sensitive telescope for X-rays created was unveiled through NASA in 1999. it remained in that position for many years. The telescope is able to study phenomena that had previously not been discovered for instance, the speedy period when the space particles disappear into the form of a black hole.

In comparison to the standard optical telescopes that we're all acquainted with, X-ray telescopes stand out. This Chandra X-ray Observatory makes use of particles with higher energy that are specifically X-rays to make images instead of relying only on light visible.

The shield-like environment of our planet stops Earth from being able to see X-rays originate from different regions in the universe. These must instead be captured by telescopes which have been placed into the orbit of Earth. Chandra is a satellite that orbits the Earth at a distance 200 times greater than Hubble's altitude is 25 % more sensitive than other telescope that uses X-rays.

It is focused on the regions that have substantial energy in the upper regions. While we are still attempting to discover the dark matter mysteries it is the Chandra X-ray telescope has provided so far crystal clear images of supernova remnants, quasars and explosions of stars and black holes that are supermassive. In addition, it has identified Pluto's low-energy light rays in the early part of in this decade.

Chandra could enhance our understanding about the origins of our universe by providing details about the creation and demise of stars. Chandra could also allow us to determine whether the planets in distant galaxies are habitable.

What is the sort of life found on other planets when you talk about life in general? Space mission Juno will introduce people to the gas gigantic Jupiter follows to our checklist of NASA successes.

THE JUNO SPACECRAFT JUPITER TRIP

The first machine to travel through the Asteroid Belt is NASA's Pioneer 10 probe in 1972. When it was completed it made history once more by becoming the first probe to see the "outer globe," in this case Jupiter.

A second probe is in the process of examining the huge gas giant. Juno is an spacecraft, which was launched on the 5th of August in 2011, at the Cape Canaveral Air Force Station. Its mission is to find out more about Jupiter particularly its gravity fields as well as the fierce atmosphere. On July 4, 2016 Juno finally came into contact with Jupiter, the massive world. Juno established a new space record while traveling. Its stunning spacecraft is driven by three 30 feet (9-meter) solar panels. Juno measured 493 millions miles (793 million kilometers) away from the sun January 13th in 2016. There's been no solar-powered spacecraft to cover this distance.

Juno is on an elliptical orbit about the gas giant. It is traveling in a direction that brings close to 2,600 miles (4,200 kilometers) of the cloud tops of Jupiter. So far, Juno has made several astounding findings. We now are aware from the information it collected that Jupiter's famous bands go beyond decoration for the surface; jet streams which power these bands can stretch at least 1 864 miles (3,000 kilometers) beneath the surface.

Two years earlier than Pioneer 10, another spacecraft did a more remarkable accomplishment. It saved the human race who would have died on the planet for ever instead of exploring the asteroid belt or study Jupiter.

APOLLO 13 MISSION CONTROL BRILLIANCE

Apollo 13 was headed to the moon. It took off on April 11 in 1970. In the spacecraft's first 55 hour and 55 mins, nearly all of the components necessary to ensure that life was maintained on the spacecraft were destroyed the explosion.

The error with the thermostats for the oxygen tank in the year 1965 was just the beginning in a series that ultimately led to the incident that caused the. After a fan was switched on, the tank No. 2, which was damaged prior to the launch was suddenly was ruptured. This was the catalyst for what would be one of the greatest teams' rescues ever. In light of the many issues that occurred on Apollo 13, the fact that the astronauts James Lovell, John Swigert as well as Fred Hayes made it home healthy and alive is an engineering feat.

A blast rattled the spacecraft just minutes after the crew had completed the television broadcast from space, informing America that all was as planned. An incident caused another. The second oxygen tank was malfunctioned due to the force created by the explosion of Tank No. 2 exploded. Three of Apollo 13's fuel cells were shut down immediately following the explosion. Additionally, they released oxygen into space despite being 200 miles (321,868 kilometers) away from its home Apollo 13's normal power source of water, electricity and oxygen as well as heat and light was disconnected.

The creative thinking that follows shows the brilliance of our human mind as well as the human spirit. Apollo 13's astronauts Apollo 13 survived on nearly nothing but water, food or sleep, all while exposed to temperatures that dropped down to a mere minus temperature in the effort to save what energy, food, water as well as oxygen were accessible. Within less than six days they suffered a loss amount of 31.5 pounds (14.3 kilograms).

Between April 11-17 The staff of NASA's Mission Control facility worked to discover a way to get these men back to their home. They estimated months of information in a matter of days. Though it was not designed to be used for this purpose but they did manage to obtain the lunar module in order to assist their crew members and bring the spacecraft back to Earth. Its lunar module's systems was not compatible with the canisters were used by the command module for removing carbon dioxide. Therefore, using the items they already had such as cardboard, plastic bags as well as adhesive, Mission Control created a solution for the team to fit them.

However, getting the car to a proper trajectory for an Earth landing was perhaps the most difficult problem of all since there were no controls, extended life support and an electronic navigation system. In the days prior to the initial explosion Apollo 13 had already started with preparations to make the moon landing.

Mission Control created a strategy. A significant star was found to serve as the foundation for Onboard Navigation. The system broke. NASA found a way to make use of the sun as opposed to an algorithmic process which typically took three months. They also discovered an approach to make use of the moon's gravitational pull in order to move the vehicle since they wanted to preserve all the energy for the return journey.

It was discovered that calculations based upon the sun's position were exact within a fraction of a degree. After orbiting the moon Apollo 13 began to descend into Earth. Its lunar modules walls been soaked with water throughout the course of those cold nights that when the spacecraft finally came on and warmed up to return there was a heavy rain in the cabin.

In April of 1970 Apollo 13 made a successfully splashdown into the Pacific. While all astronauts were uninjured however, the spacecraft, obviously, wasn't. But that was the norm in the day. Space shuttle Columbia was the first. Columbia became a landmark in 1981. It gave NASA the first reusable operational spacecraft.

A REUSABLE SPACECRAFT NAMED The SPACE SHUTTLE

The Apollo program was nearing the end of its run in 1972 and NASA was pondering its future technological plans. Rockets that were used in the Apollo missions were spacecraft that had been built one-time. Cost per mission was to put it simply massive. It's not just that a spacecraft that could be reused reduce costs and time, it also would be a significant technological advance.

NASA developed the basic concept, consisting of two rocket boosters that are connected to an orbiter module an external fuel tank along with other elements, shortly after the presidency of Richard Nixon announced his intention to create a spacecraft capable of performing a variety, possible infinity of missions.

The spacecraft faces a variety of issues. NASA required a completely innovative concept for a heat shield because the technology that was used to shield prior spacecraft from hot atmosphere was dissolved in the reentry. The company developed a way of cover the spacecraft using ceramic tiles in order that heat would be absorbed, without damaging the materials. It was also the cause of another significant change. It was essentially that the old spacecraft fell into the atmosphere and then sank into the sea. Reusing any equipment after splashing into water can be difficult. The spacecraft will be able to land on a landing strip in the same way the way a glider might.

From the start of the program until the launch of the first spacecraft, it took 9 years. The initial mission of Space Shuttle Columbia, which began in 1981, was an enormous success. NASA could create an reusable spacecraft.

Four more space shuttles--Challenger, Discovery, Atlantis, and Endeavour--arrived after Columbia. The five shuttles had completed 135 total trips Between 1981 and 2011 most of which included stays on the International Space Station.

The ISS was the product of a remarkable group effort that aims to increase space exploration. We can go back to the start of that remarkable mission to discover the very first humans to explore the other side of the moon.

A LUNAR CHRISTMAS APRIL 8 A LUNAR CHRISTMAS EVE: THE APOLLO 8 MISSION

Before NASA could launch a team on the moon, the company had to be certain that they had the technology required. The journey itself isn't easy. Only a handful of astronauts left Earth's orbit in 1968, however no of the pioneers in the early days were able to do such a thing.

In reality, the need for hardship could be an effective motivator. We'll look at the reasons in an instant, NASA was pushed by the political establishment to complete the American lunar landing before the decade was over. The time was crucial. Thus, on 19 August in 1968, the federal government announced that a manned fly-by operation would take place during the month of December.

The mission was assigned to astronauts Frank Borman, James A. Lovell as well as William Anders, who had trained for a completely different sort of mission. They underwent a intensive training prior to being put on an impressive three-mile (110.6-meter) Saturn V Rocket and then launched into space on 21 December in

1968. It was officially that of the Apollo 8 mission.

It was the day that the Apollo 8 crew arrived at the lunar orbit on 24 December following three days of waiting and one unpleasant vomiting experience. The millions of viewers who were on earth accompanied the crew in spirit. In an unprecedented television broadcast event, the Apollo 8 mission was broadcast live across the globe to viewers. Borman, Lovell, and Anders established the mood of the Christmas Eve audience by quoting in The Book of Genesis as images of the Saturn V flashed onscreen.

One of the photos that they captured would later prove to be extremely important. The famous photo that is inaccurately called "Earthrise," depicts our blue planet encased in darkness while floating above the moon's horizon. This stunning photograph has "been celebrated as having contributed to the starting and growth of environmental activism" as per NASA's official site.

The safe return of the crew on the planet they grew up on, December 27, 1968 signified the conclusion of the mission. Neil Armstrong's "one small step" onto the moon's surface was possible due to the hard work that was Apollo 8. Each new bit of data, we move closer towards the much-anticipated sequel, which will be manned missions to Mars. With regard to the red planet

A ROCKHOUND IN THE RED PLANET The Mars Science Laboratory

There's an SUV-sized nuclear powered vehicle that is located on the other side of the globe, which is at 225 million km (140 millions miles) far from us and also has its own Twitter account. Let's talk with us about Curiosity Rover Ladies and Gentlemen.

It's likely that you know what world the NASA's Mars Pathfinder initially landed on in 1997, without ever orbiting the planet. Its

Mars Science Laboratory, a $2.5 billion mission that includes an rover with six wheels, named "Curiosity" as its main draw, is one of the most ambitious successors to the Pathfinder. The earlier rovers used solar panels to power their engines that made them susceptible to Martian dust storms, as well as periods that were dark. Curiosity is able to avoid these problems because it produces electricity by using plutonium.

Curiosity is a huge Martian Rover. It's 10.4 feet (3.04 meters) long, nine inches (2.74 meters) wide and seven inches (2.13 meters) tall. It weighs 1982 pounds (899 kilograms). Comparatively to the two prior Mars exploration vehicles, especially Spirit and Opportunity it is four times more heavy and is twice the length. And, of course, the Pathfinder Rover, which looks like a microwave in dimension, is now smaller than Curiosity.

Because of the size of the craft, NASA had to carry the landing process in a challenging manner which included broken parachute, as well as used rockets. The 6th of August, 2012 on the 6th of August, 2012, it was reported that the Martian Science Laboratory successfully landed on the Mars's surface. Mars. For the past six years, scientists have been dedicated to analyzing the geology and climate of our closest planetary ally. Curiosity has found evidence suggesting that liquid water as well as organic compounds might have existed on Mars in addition to other aspects.

The bot is extremely popular on the internet. 3.94 million people are following Curiosity's Twitter account. NASA Twitter account of the social media group. account on Twitter for Curiosity.

A wealth of information on The Red Planet has been provided to us through Curiosity along with the other Mars Rovers. Many millions of Americans imagine one day in the future when NASA can send someone to space. If it wasn't in the future for another success in our list, this popular dream would not be possible at all.

The FIRST American to be in space, FREEDOM 7

Alan Shepard was the first American to go around the globe in May the 5th of May, 1961. The first American to orbit the planet wasn't to orbit the earth; this honor goes to an Soviet Cosmonaut with the name of Yuri Gagarin. Shepard nevertheless marked the first time NASA made an appearance in the history of space flight by humans.

For NASA this was an extremely stressful day. The countdown lasted over 24 hours to complete and was divided in two parts so that Shepard and his team could rest ahead of the critical moment. After several brief inspections of the equipment performed by NASA the launch team, they finally had T-15 minutes before launch. Shepard was aboard and the pilots of the launch vehicle were ready, and the launch vehicle worked properly. After that, the clouds began to form.

The weather was not a concern during the launch. The photographer who was documenting the most important NASA launch to date however, ran into a problem. NASA chose to wait until clouds clear before they launched. In the meantime for the weather to clear, one of the power inverters on the orbiter began to fail. The engineers took 86 minutes to fix the problem. The countdown recommenced. There was a delay around T-15 the second time around, as the result of NASA's decision to study the same piece of equipment for navigation two times.

The rest of the countdown was smooth The launch, which took place at 9:34 a.m. was success. Shepard had a height of 116.5 miles higher than Earth as he entered the The orbit of the Earth (187.5 km). He flew for fifteen minutes and 28 seconds on the moon, and flew around the Earth for 303 miles (487 kilometers) around Earth at 5134 miles per hour (8,262 kilometers per hour) (8,262 kilometers per hour). The pilot had completed an unfailing mission, and established the benchmark for the future NASA astronauts as he crashed into the Atlantic Ocean.

After eight years, NASA launched the mission which established its place as a major player in the world of space. Given the importance the mission was theories persist to question its authenticity in the present day.

APOLLO 11, A TRAIL on the Moon

The president John F. Kennedy revealed the mission to be NASA's most successful, only 20 days following Alan Shepard orbited the Earth: America was going to the moon. NASA began with the Apollo space program as a response to the lunar landings.

The time it took to complete Kennedy's goal The fire that erupted at the launch pad claimed the lives of three astronauts on Apollo 1 in 1967, that was an utter disaster. The mission was followed by nine more executed by NASA in the next two years, to evaluate different aspects of the technology. If delays in equipment occurred, NASA simply shifted to another equipment to keep things going at a rapid pace.

However, the first mission to be the first to step feet in the lunar surface is Apollo 11. The moon was visited by 530 million people around the world were watching with anticipation when NASA astronaut Neil Armstrong set foot on the moon's space on July 20, 1969. He uttered the phrases, "One small step for humankind and one huge leap for humanity..

The landing itself was an important event, many folks believe it was made up and it could never happen. True, it was an extravagant Hollywood spectacle that was planned and choreographed. However, it's due to the fact that it was NASA's biggest event, which was a milestone to be recorded in history, and an achievement that was impossible before the time of space.

Five more Apollo moon-related missions. Only a few dozen have been onto the moon. Every one of them was part of Apollo 17 and the Apollo program as astronauts that is NASA's continued credit. Late Gene Cernan of Apollo 17 was the final person to walk on the moon, as of the time of writing. The astronaut explained why Apollo 17 was significant prior to returning back to Earth.

He reported to Mission Control, "I'm on the surface. This gene is Gene." When I take the ultimate step of man away from the surface, and then come back on Earth for a time-- but, we're hoping, not much longer--I'd rather (say) that I believe it will be recorded in the history books: the American struggle of today has formed the future of man. When we go to the Moon and return to Earth, we leave with peace and hope to all humanity like we did when we arrived. God willing Good luck to all the Apollo 17 crew!

Milton Keynes UK
Ingram Content Group UK Ltd.
UKHW021917281024
450365UK00017B/821

9 781999 452384